# SUSTAINABLE FRESHWATER AQUACULTURE

NICK ROMANOWSKI is a biologist who has worked with aquatic plants and animals for more than thirty years, from managing a wholesale aquarium to commercial breeding and propagation. He has written extensively on wetlands and plants for habitat and water treatment, and has a particular interest in the environmental impacts of aquaculture. Nick's earlier books include a best-selling Australian guide to freshwater aquaculture, as well as guides to wetland creation and planting, and various other plant groups. He has also written on various aspects of aquaculture for many magazines and journals, including *Earth Garden*, *Grassroots* and *Permaculture Edge*.

A UNSW Press book

Published by
University of New South Wales Press Ltd
University of New South Wales
Sydney NSW 2052
AUSTRALIA
http://www.unswpress.com.au


First published 2007



National Library of Australia
Cataloguing-in-Publication entry

Romanowski, Nick, 1954– .
Sustainable freshwater aquaculture: the complete guide from backyard to investor.

Includes index.
ISBN 0 86840 835 2.

1. Aquaculture - Australia. 2. Aquaculture industry - Australia. I. Title.

639.310994

Design Ruth Pidd
Cover Photo of barramundi by the author
Printer Everbest

# SUSTAINABLE FRESHWATER AQUACULTURE

## The complete guide from backyard to investor

Nick Romanowski

# Contents

# Contents

# Acknowledgments

This book could not have been written without help, advice, commentary and diverse other assistance from many people over the course of the past three decades. It is impossible to remember who they all are after such a time, and there would be far too many to credit even in just a few paragraphs. However, I would particularly like to acknowledge the various people who have helped during the actual writing of this book, from allowing me access to their properties for photography, to information and discussions on a wide range of issues and methods.

My thanks to Stan Armistead, Glenn Briggs, Laurence Gedye, David Holmgren, Dave Keenan, Bill Lambert, Rob McCormack, Hugh Meggitt, John Mosig, Kathy and Bruce Newell, Noel Penfold, Bill Pomorin, Heinz Staude and Ross Rickard. I'd like to specially acknowledge the journal *Austasia Aquaculture*, an important information source for anyone who wants to keep up with what is happening with aquaculture in this country. Although most of the photography and illustrations are my own, I would like to thank Bruce Sambell of Ausyfish for allowing me to include his photo of sleepy cod. Thanks also to various people at UNSW Press, particularly John Elliot, Di Quick and freelance editor Brendan Atkins.

Get hooked on fresh fish
5.99
2.99

# Preface: Towards sustainability

*Most of the fish sold worldwide are still taken from the wild, but the harvest is far in excess of what can be sustained, and aquaculture must inevitably become an increasingly important source of seafoods.*

Worldwide, wild fish stocks are vanishing through overexploitation, pollution and some possibly also through climate change, with most of the world's major fisheries stretched to the limit or even in decline. Though Australia is better off in many ways than more-overpopulated countries, the writing is on the wall: the old hunter–gatherer ways of taking what you want from wild stocks are over. Instead, the emphasis has shifted towards cultivating an ever-increasing number of the aquatic foods we actually need.

Aquaculture is the art and science of raising living organisms in water (whether these are plants or animals, or even some types of bacteria) for food or profit, and sometimes even just for the pleasure of it. It is an art because it takes a considerable degree of intuitive skill to do well, and a science because its practitioners need to understand the basics of aquatic biology, as well as the chemistry and physics of the underwater world, to learn the art properly.

Over the past two decades, aquaculture worldwide has been growing at an incredible rate and, though few of the most productive species raised overseas would be welcome here because of their seriously invasive natures, Australia has not lagged too far behind, with an overall growth of around 15 per cent per annum since my earlier book *Farming in ponds and dams* was published. Although this was the best-selling aquaculture book of the 1990s, so much has changed in the past decade that only a completely new book could do justice to the new species, directions and environmental concerns raised by aquaculture today.

## *Marine aquaculture*

This book concentrates on freshwater (or to be more accurate, inland) aquaculture because freshwater and marine aquaculture are two different worlds. The water chemistry is very different, and the biology behind raising marine organisms is much more poorly understood in most cases, to the extent that very few species have been bred successfully over more than a generation or two. Marine aquaculture is also an expensive proposition for well-financed companies and corporations, and in many cases it can best be described as serious gambling, given our present state of knowledge.

Even the process of getting a permit to pump seawater to on-shore facilities requires large-scale financing, as does finding suitable (invariably expensive) coastal land. The planning process can take years: environmental impact statements, elaborate leasing arrangements, and restrictions on how and what you can breed and sell. Part of the reason for these restrictions is the high fail rate of marine aquacultures in the past, which has made banks increasingly reluctant to fund such ventures, and also the environmental damage both failures and successes have often left behind.

If you decide on cage aquaculture out at sea instead, your problems will multiply. Permits will be still harder to come by: the seas are public property, and many government organisations and councils, not to mention people who may live in the area or visit regularly, object to the visual and physical pollution that has been associated with many of these systems. The cage must be strong to survive storms, be massively anchored, and be made of expensive corrosion-resistant materials, though even then it won't last all that many years. Most marine animals are difficult to breed and then raise, so you will need access to suitable biologists just to deal with this aspect alone, and many species are almost impossible given our present knowledge. For healthy growth, their young often require very specific live menus that may change as they pass through different growth stages, and that we also know little about. Though some fry may accept substitute foods, they will often grow poorly or abnormally as a result, and such fish don't look as saleable as their wild relatives, or taste as good.

Many marine aquacultures have yet to make a profit, although to be fair these speculative ventures are doing pioneering work in completely new fields. The sole supplier of breeding stock for the Australian prawn aquaculture industry recently commented in a trade journal that $200 million had been spent on infrastructure to produce $50 million of prawns. This in an industry which is still completely reliant on trawlers to catch *all* of its breeding stock, because they have been unable to produce healthy adult spawners in captivity!

Prawn farms, as well as various other marine aquacultures such as mud crabs, also have image problems, because these conspicuous developments are often carved out of mangrove stands, which are an all-too-depleted breeding ground and nursery essential for the continued health of wild fisheries. Nor is it easy to regenerate cleared mangroves if a farm fails, as the carving out of ponds often exposes acid sulfate soils, which few plants will grow in. Finfish marine aquacultures also affect wild fisheries, taking stock from the wild to fatten in cages, or having unhealthy snapper escape into the wild, blackened by exposure to the sun in waters much shallower than their preferred depths.

Many of these aquacultures are also associated with pollution problems, from waste food and excrement build-up on seafloors, to toxic chemicals from anti-fouling paints. And then there are public health questions as seacaged salmon are fed canthaxanthins to turn their greyish, poor quality flesh to a healthy-looking orange – even though the use of these food colourings in processed foods is restricted in many parts of the world.

In short, there is good reason to think that most of marine aquaculture is far from sustainable at present, and that many of these emerging industries are dependent on clean, well-stocked seas for their livelihood, even though they continue to cause localised environmental damage.

Take the case of southern bluefin tuna, the great success story of Australian aquaculture in recent years, which has been responsible for most of the recent overall growth of the industry in the past few years.

Here we have an industry which takes all of its stock from the wild, and resents the harvesting limits which have brought further expansion to a halt – even though these limits have been set so the wild resource will hopefully not be further depleted. The seacages originally used were deemed to be too polluting for the sea floor below, so this problem has been 'solved' by anchoring them further out to sea. And 10 kilograms of wild-caught fish (mainly pilchards) must be dragged from the sea, with considerable wastage of fuel and wild resources which underpin other fisheries, to feed and produce each kilogram of 'value-added' tuna.

This isn't farming in any real sense, but a feedlot operation which is *totally* dependent on already overexploited seas and considerable usage of fossil fuels for its very existence. Despite these problems, the bluefin industry continues to reap significant profit, little of which has yet been invested in finding substitute feeds, or in breeding the species rather than depending on wild populations, or in making any provision for the removal and reuse of the wastes it produces.

There are some sustainable marine aquacultures already (for example, oyster farming), though these may also be dependent on wild stocks harvested judiciously. They will be joined by others as developers come to realise that sustainability is not just a buzzword, but a fundamental requirement in aquaculture, as it must be for the long-term future of any other type of agriculture. For the moment, however, too many marine grow-out operations continue polluting and carrying on with their outmoded practices while trying to put a positive spin on the all-too-obvious problems they generate instead of doing something about correcting them.

## *Freshwater aquaculture*

By contrast, there are two levels at which freshwater aquaculture could be used to produce increasing quantities of fish and crayfish, using only proven technologies and the biological resources already available. At the most fundamental level, few farm dams produce anything but muddy water for livestock, yet many of them are suitable for raising at least one or two species of fishes or crayfish and, if most of the dam is fenced off from livestock, could also grow a wide range of edible and often very productive plants.

If the estimated half-million dams in Australia were stocked with suitable species, they could add considerably to the fresh fish and crayfish presently eaten in this country, particularly in more isolated areas where fresh seafoods are unobtainable. But you can also start with freshwater aquaculture in a small way with a large pond in the garden, or learn the skills you need to breed and raise fishes and other freshwater animals in a few aquariums, while making a profit.

If you're only raising a few fish for yourself, you won't need special permits for stocking them in most places. Many of the fishes and all of the crayfish available will breed by themselves in suitable conditions, or with a little help, as described later in this book. More is known about their dietary requirements including suitable live and prepared foods than for the vast majority of marine fishes, and good quality prepared foods are now available for most of the freshwater species that have been farmed for some time.

A backyard pond can be topped up with fresh water from a tap in most cities (though you may need to treat this to remove chlorine and sometimes other compounds before use), and once full won't use any more water than a lawn of the same area takes to keep it green, while dams fill naturally with the rain that falls from the heavens. Ponds and dams can be kept clean with a combination of aeration, judicious plantings (themselves a possible source of food or profit) and reasonable stocking rates, so that any runoff from dams during wet periods should be of good enough quality to be almost drinkable.

If the dam or pond is large enough, or the water filtered through plant beds, you can swim in it, harvest crayfish and fish from it, plant it with edible and ornamental species according to taste, and build a jetty and sun canopy complete with hammock. Freshwater aquacultures are more than just something we can fit into our everyday lives – they can become its living epicentre.

On a commercial scale, successes to date have been mostly moderate, but these have set a clear and well-defined framework for a much larger freshwater industry than was ever dreamed of a decade ago. It is now possible to produce good quality food from larger water bodies almost anywhere in Australia, even in fairly saline or alkaline inland waters, as many of our native fishes and crayfish are well adapted to these harsh environments. Some inland saline waters, with a little modification, are even suitable for farming hardier marine fishes.

Already, commercial quantities of some fish are being harvested and sold from reservoirs in various parts of Australia, with others showing promise in experimental trials and still others which have barely been assessed so far. Particularly interesting is the increasing use of cage or raceway culture systems in larger water bodies, where there has been an emphasis on economical ways to remove the solid wastes which cause many water quality problems – something most of marine aquaculture has signally neglected.

Droughts, increasing populations, over-allocation of present resources and the need for greater environmental flows to waterways in recent years have put an end to the unrestricted use of water, and there is going to be even less available in the future. Even in major cities the trend is towards making mains water more expensive, encouraging efficient use and reuse, discouraging waste, and harvesting rainwater from your own property, rather than putting in more dams on already overextended rivers.

Yet too many intensive aquaculture systems continue to dump the considerable volumes of water used to flush their ponds on pointless activities such as keeping pasture or lawn green. Waste water would be better used to irrigate plantations, market gardens or other crops instead, or should perhaps be cleaned up through created wetlands.

Water treatment wetlands are already a requirement for many new developments which use considerable volumes of water, and can also be used to clean aquaculture waste water to the extent that it can be reused rather than disposed of. If the wetlands are extensive enough, water harvest and recycling could become a more-or-less closed system.

In the wider world, integration along these lines is already being practised by some far-sighted growers, usually on large properties where any water allocation is first run through aquaculture ponds, then carries its light burden of animal wastes to irrigate and fertilise crops ranging from annual vegetables to trees. Even the sludge from the bottom of an aquaculture pond can be harvested from time to time, and safely used as a high-nutrient mulch or compost on terrestrial beds.

The use of chemicals (whether toxic or otherwise) is already minimal in much of freshwater aquaculture, mainly because the first living creatures to feel the impacts would be the aquatic animals themselves, many of which are extremely sensitive to even quite mild toxins. As a result, cultured foods from freshwaters are among the cleanest available, yet already some growers are looking at ways to improve this still further, including feeding with only organically raised feeds from certified properties.

Many of these ideas and developments are still very new to the aquaculture world, and most growers would argue that they are only the icing on the cake of an industry which has had to grow up in an era of increasing awareness of environmental degradation. Yet as a result much of freshwater aquaculture is already close to sustainable, and with more people growing food in their ponds and dams, it could become not only a considerable source of clean, high quality food, but also a further step towards a broader ethos of sustainability.

Though this book is primarily intended to be a complete, basic guide to freshwater aquaculture, from constructing ponds and dams to harvesting the crop, wherever possible I have emphasised the extra steps that could be taken towards true sustainability. What pleases me most about these is that they are relatively few compared to the changes needed in most other primary industries, and they don't necessarily add much to the cost of existing processes. It would not take many changes to make freshwater aquaculture the healthiest primary industry in a country that already supposedly prides itself on its clean and green image.

## *Reading further*

All too often, I meet people who have just surged into aquaculture with half-baked ideas drawn from a newspaper article or some fringe notion from the Internet, and have proceeded as if all the centuries of human experience of raising aquatic animals worldwide is now redundant. Almost invariably, these people throw away their money on impractical schemes – sometimes just a few hundred dollars, but in many cases figures in the tens of thousands.

The biology of living things has not changed in the computer age, nor have the underlying tenets of aquaculture, and though common sense applies in this field as much as in any other, the ground rules are very different to those for any other type of farming. Getting a grasp of the basics of aquaculture through reading, research and small-scale experimentation is the essential first step to successful aquaculture.

Although the information in this book is complete enough that you shouldn't need to refer to other sources to understand how freshwater aquaculture works, anyone with a serious interest should treat it as a springboard to further and more specialised reading and research. For this reason, each chapter concludes with a brief summary of particularly useful books and sometimes articles which will extend the ideas already discussed.

Few aquaculture websites include information that is directly relevant to the way the chapters of this book are set out. Some sites include useful details of current research and experimentation so they are included below rather than in specific chapters. Much of the general information about, for example, practical aspects of aquaculture, available from various state government department websites, has often not changed much in recent years, and so is not included for this reason.

The following sites may be worth checking, especially if you are interested in commercial aquaculture, but keep in mind that claims made on some sites may be exaggerated or distorted, or sometimes even seriously wacky! Commercial sites aren't included (even though some are worth checking) and any claims which seem even slightly doubtful should be verified against independent sources before taking advice or buying from any website.

Wiith these caveats in mind, an obvious starting point would be the Australian Aquaculture Portal at <http://www.australian-aquacultureportal.com>. This site is managed by the Australian Aquaculture Council and provides links to news updates, articles, industry groups, research, and commercial products and services.

The CSIRO website provides a useful overview of its current and recent research programs in Australian aquaculture at <http://www.marine.csiro.au/research/aquaculture>, but it focuses mainly on marine species and issues, rather than inland or freshwater aquaculture.

For more detail on the problems associated with marine aquaculture, the World Wildlife Fund (WWF) has an ongoing global campaign highlighting some of these issues. Visit the WWF website <http://www.panda.org> and search for 'aquaculture' for more information. The Australian Marine Conservation Society at <http://www.amcs.org.au/campaigns> provides an Australian perspective on sustainable fisheries and aquaculture.

*Plant ponds with EPDM liners at left, and freestanding fibreglass ponds to the right.*

- Making small ponds
- Planning for earth pond and dam construction
- Machinery for pond and dam construction
- Making earthen dams or ponds
- Other dam-lining materials
- Commercial ponds
- Moving water
- Modifying existing dams for aquaculture
- Reading further

# 1 Holding and moving water

To start with aquaculture, you need water and something to keep it in. Although obvious, this basic need remains one of the most neglected aspects of the whole field, to the extent that most books barely even touch on the subject. Not surprisingly, the lack of information leads to many serious problems for people planning to construct ponds and dams.

Problems are much less common when setting up a commercial operation with many earth-lined ponds, because the sites for such enterprises are usually chosen to be as flat as possible, and are almost invariably on heavy clay soils that will hold water well, even without compaction. On a smaller scale, ponds and dams are readily fitted into the best sites available on properties used for other purposes.

Whatever type of pond is being constructed, a reasonably flat site is easiest to build on because earth-moving costs increase exponentially as slope increases, while the volume of water that can be retained per dollar spent decreases exponentially. It may be possible to build a pond on a steeply sloping property, but the resulting pond will either be very small and narrow, or cost tens of thousands of dollars to construct, and has an increasing risk of failure as the walls climb higher. Completely flat sites are ideal for building relatively shallow ponds designed specifically for aquaculture, but are not cost-effective for deeper water storage reservoirs – and shallow water storages are subject to excessive evaporation.

Apart from considering the location, it is essential to check what laws and by-laws apply in your state and local area. These have been changing rapidly throughout Australia for more than a decade now, and will undoubtedly continue to do so over the lifespan of this book, so it is impossible to give more than general guidelines here.

For smaller backyard ponds, especially those fed by roof runoff, you may need nothing more than a planning permit through the local council or shire, though it is likely that the pond would need to be fenced to keep smaller children away as is required for backyard swimming pools. Larger dams and aquaculture farms are also often completely fenced, not only for safety reasons but also to deter poachers, and to prevent livestock from muddying the water.

Larger ponds, anything from a small dam upwards in size, are covered by a wider set of rules and may require permits at several levels. For example, in Victoria new dams for any commercial purpose need to be licensed by one of the recently formed rural water and catchment authorities which regulate the proliferation of dams that have reduced environmental flows in many catchments. Other states and territories are likely to have similar regulatory bodies in future.

There are also increasingly stringent controls on where dams can be built, particularly in or near watercourses both permanent and ephemeral, and where water may back up onto adjacent properties. The environmental officer of your local shire or council is often the best source of contact details for all authorities which have a say in the construction of new dams and larger ponds in your immediate area.

## *Making small ponds*

For aquaculture purposes, any pond which holds less than 100 000 litres of water would be considered small, though with proper design and management it could still produce a reasonable amount of food. It is difficult to make earth-lined dams of this size, though not impossible if the soil is a heavy, good quality clay that will hold water without much compaction.

Small puddled-clay ponds have been built by hand in the past, though mainly in countries with regular summer rainfall. These are only shallow affairs, lined with wetted clay and sometimes animal manures, packed down to form a watertight layer. Such relatively thin clay layers don't hold water as well in deeper ponds, as water pressure and seepage increase with increasing depth. In Australia, a more serious problem is that in most places a puddled-clay pond will dry out and crack over summer, after which it will no longer hold water.

If the earth-lined pond is large enough, it can be made by machine using the methods discussed in the next two sections for larger dams. Even a bobcat (carrying heavy loads of clay in a scoop) will compact the soil sufficiently to hold water in a shallow pond up to perhaps 1.5 metres deep, but heavier machinery or a roller of some kind will be needed for a good seal at greater depths.

Most small ponds are made with pond liners, less frequently with fibreglass, ferrocement or concrete. Pond liners are the easiest to use and install, the entire site being dug out, levelled and then lined with protective materials such as wads of wet newspaper before the liner is placed. There are many grades and qualities of pond liner, the suppliers of which provide booklets showing how the ponds are made, how the liner size should be calculated, and giving information on the expected lifespan of the product.

The cost of the liner is usually a good indication of how long it will last, the lightest and cheapest materials often becoming brittle from exposure to ultraviolet within a few years, while more expensive products such as butylene and EPDM (ethylene propylene diene monomer) will last for many decades and have warranties to match. These heavier materials are also more elastic than thinner liners, and don't produce toxic gases, as some cheap liners do – extremely undesirable for anything you are planning to eat later from the pond!

The main disadvantage of liners of any kind is that they are relatively easily to puncture if carelessly handled or walked on, and that it is often difficult to locate the hole for repair work. Used on a large scale, liners are usually protected with a covering of soil or sand, and this can also be done in a small pond, though the exposed sides will need some other protective lining over their surface.

Fibreglass is mainly used for smaller ponds as it is too expensive for larger works, and it can seriously damage the health of people exposed

▲ *A sheepfoot roller in the foreground and a scraper behind, essential tools for compacting marginal clay soils.*

to the fumes produced during construction. The advantage of the heavier grades of fibreglass is that they are strong and will last for long periods even in sunlight, and can even be made portable and free-standing if required. However, the water temperature of ponds standing on the ground rather than sunken is more variable, so above-ground fibreglass ponds are best used for plants rather than animals.

A concrete pond is probably the best and longest lasting type, but can be ruined permanently through careless construction, and may be expensive if built by a contractor. It should be poured in one single and continuous process so all joins and seams bond before setting, preferably with the aid of a team of concreters experienced in swimming pool manufacture.

All linings, reinforcement, drainage pipes and formwork must be in place before the pour, during which the concrete is brought in by trucks. A concrete vibrator should be used to settle the mix within the formwork of the walls to remove all air pockets, but it must be used only briefly or the sand may separate out and weaken the mix.

Small ponds can be made with more expensive ferrocement mixes, half sand and half very fresh cement mixed together – it does not hurt to add one of the many cement waterproofing compounds available as well. The reinforcement for these can be as light as chicken mesh, folded and formed over a liner of builder's plastic. However, the pour must once again be a continuous process, so several people may be needed to lay and smooth the mix while one person makes it up in a mixer.

All concrete and cement mixes must be cured after setting, to remove free lime from the surface layers exposed to the water. Painting with cleaning-strength acetic acid (vinegar), then filling the pond and allowing it to stand for a week or two will remove much of the lime, but the process may need to be repeated for best results.

Concrete livestock troughs can also be used as a ready-made small pond, and have the advantage of being relatively portable. The smallest sizes are too small and awkward to be much use, but those holding at least 5000 litres, and having a diameter of 3 metres or more, can be used for many purposes, from breeding ponds to nurseries for young fish or crayfish. They must be set on a level, raked bed of sand or fine gravel to evenly spread their weight (around 7 tonnes filled!), and the inside needs to be cured as described above.

In the backyard, one of the most commonly used types of small pond is the swimming pool, which can usually be adapted to a natural filtration regime by using the existing parts, but without the use of chlorine. Systems can also be made up to biologically filter such pools, and if the water volume is large enough and not overstocked, they can even continue to be used as 'natural swimming pools'. Filter systems including those which can be adapted to swimming pools are looked at in more detail in chapter 3.

▶ *A leaking dam in uncompacted dispersive clay in Western Australia – the line above which the water level never rises is clearly marked by a 20-year-old pine tree!*

# *Planning for earth pond and dam construction*

## Sites for ponds and dams

Choosing sites for a pond or dam is largely a matter of commonsense. The soil on the site must be capable of holding water, and as this is a major consideration it is discussed separately (see next section). There needs to be a reliable source of water for at least part of the year, whether from a catchment, bore or pumping allocation from a river. Steeper slopes should be avoided as earthmoving costs increase exponentially with gradient, and they are also harder to climb if anyone slips into the water.

A dam needs a substantial runoff zone or catchment above it in order to fill, and the drier the area, the larger this needs to be to provide a reasonable volume of runoff during wet seasons. For this reason, most dams are built on creeks or gullies where the running water tends to channel naturally, but larger and more permanent watercourses should be avoided as they can flood dramatically, breaching the wall. Increasingly, there are also planning restrictions on building anything on or near waterways in some states.

On sites with less reliable flows, the catchment can be effectively increased by forming shallow drains across the gradient to either side, so that more of the runoff from above is channelled towards the dam instead of simply draining past on either side.

Upstream water quality can be an issue, as runoff from fertilisers and agricultural poisons can be lethal in an aquaculture system. Where excessive nutrients are a problem, a series of shallow ditches or swales across the path of the runoff will slow the water so more of it soaks into the ground. While the dissolved nutrients will be used by terrestrial plants in the swale rather than by algae in the ponds, the greater infiltration will also reduce the amount of runoff reaching the dam.

Agricultural poisons are a more serious problem, and upstream neighbours producing any significant waste of this kind may make it impossible to raise anything successfully in water which has passed directly from their properties to yours. Wetlands for waste water treatment can be used to reduce nutrients and some other chemicals before the runoff reaches the aquaculture ponds, and these are discussed in chapter 3.

On flat land, which is the ideal for any larger-scale operation with many ponds, construction work is basically a matter of digging into the ground. Either stockpile the excavated soil in mounds out of the way, or use it to build up walls and driveways around the ponds and dams. On sloping ground, wherever possible, all earthmoving should be done between shoulders of higher ground to build up retaining walls, following the natural contours of the land.

## Materials

Most larger aquaculture ponds are earth-lined, made by the same methods as earthen dams – and one in every three earth-lined dams in Australia leaks. The reason for this is simple: the majority of earthmoving contractors don't know how to make a dam properly. In extreme cases I have seen dams with walls made by guesswork instead of with a level, dams with overflows higher than the lowest part of their wall, and even walls with entire trees sticking up through them! In most cases problems are less obvious, yet even many apparently well-made dams still won't hold water past a certain level.

The most common problem is where no attempt has been made to test the soil for permeability, to see if it needs compaction. Many soils which are apparently clay-rich are anything but, and limestone marls (which look like a thick, grey clay) won't hold water no matter what you do to them.

The simplest and most widely recommended test for the water-retaining qualities of a clay soil is to roll a moist sample into a long cylinder the thickness of a pencil, then gently curve this into a tight U-shape. Even this test may fail to detect dispersive clays, which can have just enough flexibility to pass the test but won't hold water unless compacted. When wetted, dispersive clays can separate into very fine particles, so small that even the slightest current will keep them suspended indefinitely – and there is always a gentle current in even the smallest earthen dam.

Such clays are the reason that many dams remain cloudy even after many years, though their particles can be made to flocculate (that is, bond together to form larger, heavier fragments which sink more readily) by adding lime. A simple test for dispersive clays is to add a pinch of lime to one of two otherwise identical jars filled with clayey water. The jar with lime in it will clear within a day or two if the clay is dispersive, while the other will take much longer or not settle at all. On a larger scale, lime can also be added to an existing dam with permanently cloudy water to help clear it, though the results may not be permanent.

Compaction with a sheepfoot roller can make even dispersive clay soils hold water very effectively, as described in the next section. If left uncompacted, dams in such soil may hold water up to the original soil level, but where the wall rises above this level they generally leak like a sieve, indicated by various water's edge plants appearing along this line with time. The dam may fill to the brim after heavy rains, but will rapidly return to the previous level, sometimes within just hours. Compaction of the entire inside of the wall with a roller starting below this waterline will sometimes repair the problem, though not tying the wall to the soil below gives a similar result, and compaction cannot fix this.

Whatever material is used for an earthen dam, there should ideally be an abundance of it, and core samples should be taken over the whole site to check this. An earth auger of diameter 10 centimetres is all that is needed for core sampling, and although this will only reach to around 90 centimetres down, the head can be unscrewed and the shaft extended with further sections of steel pipe of the same thread. Two extensions, each 90 centimetres long, will allow samples to be taken up to 2.5 metres deep, which should be enough testing for most aquaculture ponds and dams.

## *Machinery for pond and dam construction*

There are several very different types of machine which are widely used in constructing larger ponds, each of which has its uses and limitations.

Most limited is the **excavator**, which is largely a digging tool that cannot make watertight walls except in the heaviest clay soils. Many of the most crudely made failed dams I have seen were made by excavators, which should really only be used to dig water-retaining holes in flat ground. Some excavator operators claim to be able to compact dam

*▲ A 10-centimetre earth auger with a single pipe extension for sampling clay up to 1.6 metres deep.*

*▲ A bulldozer showing the tines used to rip and loosen compacted soils.*

*◀ An excavator, useful mainly for digging ponds into the ground rather than raising walls.*

walls by pounding them with the digging bucket, but this hit-and-miss proposition should only be considered if no other compacting device is available within hundreds of kilometres.

The **bulldozer** is the most widely used and versatile pond-making machine, due to its combination of power and manoeuvrability even on quite steep slopes. This capacity for work on difficult sites is made possible by the caterpillar tread, which spreads the weight of the machine while increasing traction. It also means that a bulldozer cannot compact marginal soil types, so it must often be used in conjunction with some type of roller.

A **scraper** is basically a specialised type of truck, with a scoop on its 'underbelly' which can be lowered to scrape up a layer of soil. Because of its length, a scraper needs a fair amount of room on relatively flat surfaces to manoeuvre, so it is best suited to the construction of long or large dams. As it runs on tyres with a relatively small surface area in contact with the soil, and carries many tonnes of weight when filled, a scraper does a good job of compacting most soils as it works its way repeatedly over them.

A **scoop** is a smaller version of the rear part of a scraper, but is usually towed by a bulldozer to increase soil-carrying and transferring capacity. Although it also runs on tyres, the weight it can carry is considerably less than the capacity of a scraper, so the compaction factor is reduced.

A **sheepfoot roller** (also called a calfsfoot roller) is one of the most important yet under-used pieces of dam-making equipment. The non-motorised type can be towed by a four-wheel-drive tractor or bulldozer on steeper sites, while self-propelled, vibrating rollers are more efficient where there is a large area of gently sloping ground to compact. As a roller moves over uncompacted ground, its numerous tines sink in for the first few passes, and ride higher with each pass until eventually only the tips penetrate the soil after six to eight passes.

By this stage, the soil below will be compacted for around 30 centimetres, and should be able to hold the pressure of up to 3 metres of water above. For deeper dams, an additional layer of compacted clay should be added for each additional 3 metres of water depth. In theory, even soil with as little as 20 per cent good quality clay can be compacted to hold water, assuming the clay is evenly mixed through the soil.

## *Making earthen dams or ponds*

There are some machinery operators who know how to construct earthen dams and have the right equipment to do the job properly, but they are surprisingly few. Anyone planning to put in an earth-lined dam or pond must therefore have a reasonable basic understanding of what needs to be done, as described here, and discuss this in advance with the people who are going to be doing the work. If you are told that all these fancy theories about dam construction are irrelevant to the job on hand, and that the operator has made many hundreds of dams without the slightest problem, ask to see proof in the form of the last ten dams constructed – not just those which were successful!

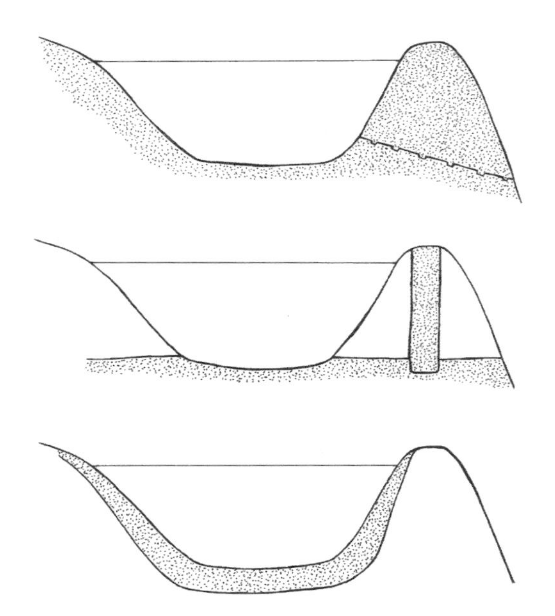

*Three dam profiles showing typical construction techniques. Top: a wall made entirely of good quality water-retaining soil where this is abundant. Centre: a clay core through the centre of the wall. Below: a clay skin used to line more porous materials.*

You should also plan to be present for as much as possible of the construction process to deal with unexpected problems, and to make sure that the operator doesn't take any short cuts. This is all too common: if the operator doesn't believe the methods you want followed are worth worrying about, he may decide to save you some money by ignoring them. It is also important to make sure the person you discuss the job with will actually be operating the machinery, rather than just instructing the operator.

Whatever type of dam or pond is being constructed, the first stage is to scrape the topsoil off the site and stockpile it for later use. Once the lining material is exposed, construction can begin, using different methods depending on the nature of the site and the amount of high quality lining material available.

If the site lacks a good supply of clay, what little there is will usually be used to line the pond. Here, the outside wall is built up of any reasonably solid material available – even sandy soil will do, though this tends to settle with time causing the wall to shift a bit before stabilising. The available clay is then spread evenly across the inside of the wall to a minimum depth of 60 centimetres, though up to a metre would be better if there is enough of it. A thin skin or blanket of clay like this *must* be compacted to hold water reliably.

Another way to deal with a scarcity of good clay is to make a vertical column of it in the centre of the wall, building up the inside and the outside around it with poorer soils as the wall goes up. This central column should not start at ground level, but should begin in a trench cut into the clay base as a foundation. As it is built up each layer of the central column should be compacted at intervals of 30 to 50 centimetres.

Where clay is abundant, the entire wall may be built up from it, and care should be taken to tie (or key) the base of the wall together with the material of the wall itself using, for example, bulldozer tines or rippers to tear up several furrows through the clay foundation to around 20 centimetres before starting to construct a wall above. Without such keying, there is a chance that water would seep between base and wall or even cause a slippage here so the whole wall tears away.

For most types of wall, the entire inside should be rolled to compact it, as described earlier, though there are some areas where the local clay soils are so heavy that even a bulldozer alone will be enough to build a wall that holds water well enough. This can best be judged by looking at dams on neighbouring properties, and discussing any problems their owners have had with them. In a deep water-supply dam, several layers of compacted clay may be needed, one for each additional 3 metres depth.

Once the wall is finished, an overflow (sometimes called a bywash) is cut into one side of it, to allow water to escape during peak flows without going over the wall. The width and depth of the overflow depends on how much rain is likely to fall during extreme flood periods. If the dam wall is to be used for traffic, a set of concrete pipes of appropriate diameter can be laid in the overflow and backfilled with lining material before compacting. Such pipes can also easily be covered with mesh to

▲ *A scraper and two bulldozers, one towing a scoop, during construction of a large water-storage dam.*

stop smaller fish escaping, or deter eels on their upstream migration. The slope below should either be wide and well-bound together with plants on a shallow slope, or lined with coarse rock rubble on steeper sites to prevent erosion.

Use of a level is essential to finish the horizontal wall, before the overflow is cut in. Although it may seem unbelievable, there are still operators around who insist that their eye is so good that they can finish a job perfectly by sight alone. Work done by such experts is distinguished by being noticeably wonky, and in some cases I have seen the overflow placed at a higher level than the *lowest* part of the dam wall, rendering the overflow redundant. Note the emphasis here – there shouldn't *be* a low point on a wall!

Finally, the topsoil can be replaced. It is often recommended that this should be spread evenly over the floor and sides of the dam, but it will have little use in deeper water. It should mainly be spread over the top of the wall and in other places to be planted. In shallow commercial-type ponds, the soil can be spread evenly everywhere, as the entire pond is usually planted with grasses and other low-growing plants before flooding.

If you plan to grow aquatic plants as shelter for animals or as crops, you will probably want to incorporate planting shelves along the sides, and these can be covered with specific depths of soil as explained in chapter 12. Even in this case there is no point in spreading topsoil in areas which will be over a metre deep, as these depths are unlikely to ever see enough light to allow plants to thrive.

▶ *Dark-brown topsoil spread on a completed dam wall, but not inside where it would have no purpose.*

▶ *A self-propelled sheepfoot roller for compacting gentler slopes.*

## Other dam-lining materials

If there is no good source of water-retaining soil on a property, and no chance of trucking enough for a clay liner from nearby, it is still possible to make ponds, even in sandy soil, by using a waterproof liner. There are diverse types of dam liners available in different parts of Australia though they are a relatively expensive option. Most are at least reasonably stable in ultraviolet light, which breaks down many plastics fairly

▶ *Bulldozer and sheepfoot roller tracks showing how much deeper the roller sinks into this already fully compacted ground.*

quickly, and of course only food-grade or guaranteed non-toxic types are suitable for aquaculture.

Butylene and EPDM membranes (discussed earlier) are the top of the range but also by far the most expensive option. Both types can be sealed or welded into very large continuous sheets, but EPDM is generally the cheaper of the two and, as it can be joined on-site with a waterproof, two-sided tape, is the most convenient to use. Other dam liners are less expensive, but also not as long-lived in sunlight. Those which are tough enough to run a machine with tyres over can be covered with a layer of soil up to 15 centimetres deep to block ultraviolet. Sandy soils are not suitable for this purpose as they erode readily during heavy rains if exposed.

The fine, powdered clay called bentonite expands dramatically when wetted (between 10 and 17 times in volume depending on type), and has been used to attempt to seal leaking dams by bombarding the water with the powder. In theory, the small particles are drawn into the leaking areas and expand there, cutting off flow. In practice, most of the bentonite expands before it even reaches the bottom, though it may still make some difference. However, a 15-centimetre blanket of dry bentonite can be used as a pond liner, if covered with a layer of soil at least 20 centimetres deep. As the pond fills, the bentonite expands beneath, sealing it. The protective layer needs to be a fairly heavy soil type to prevent erosion, which would otherwise expose the bentonite.

New types of geofabrics using bentonite are now being used to seal even toxic waste sites from below, and are useful though still fairly expensive for making ponds. The bentonite is packed into a fabric lattice, which comes in large rolls that need to be handled by an excavator with a special spreader bar attached. These are overlapped by 30 centimetres as they are unrolled, and are also covered with soil to protect the liner. None of these relatively expensive options are likely to find much use in commercial-scale aquaculture, as it more economical to find a block of land with suitable soils for ponds in the first place.

## *Commercial ponds*

Most existing farm dams can be used for raising at least some types of animals and plants, though they may need some retrofitting and cleaning up for best results, as explained in a later section of this chapter. These are designed primarily as water storages, and any aquaculture crops raised in them are an incidental benefit.

When a pond is intended primarily for aquaculture, particularly for commercial harvests, its design and set-up will be very different from a dam. Such ponds are often relatively shallow so they will warm up more rapidly, and they are usually a standardised shape for convenience in harvesting. In many cases, the harvest will be carried out by draining the pond, so the floor may slope gently to a shallow sump with an underground pipe controlled by a valve.

Laser grading of the floor is desirable though it adds to construction costs, allowing more even drainage. Grading encourages freshwater crayfish especially to spread out over the whole floor rather than instinctively congregate in deeper pools, channels and other depressions which are less accessible to predatory birds. Deep sumps are also undesirable for the same reason.

Most commercial ponds are rectangular in shape, with broad bunds between them allowing heavy machinery up to the size of a tractor to move alongside. The larger the area of a pond, the harder it is to manage and harvest. On the other hand, the smaller the pond, the more temperatures will vary between day and night, and water quality is likely to fluctuate more rapidly if anything goes wrong.

For these reasons, grow-out ponds (where animals are raised to a marketable size) on commercial farms may hold from 0.5 to 5.0 megalitres, and may vary in depth from around one to two metres depending on the species being cultured. Breeding ponds hold a much smaller number of larger animals, and are almost always smaller, with greater control over water levels sometimes being required. The water level in such ponds can easily be controlled through the outlet valve, so they can be used full for raising larger fish, or kept relatively shallow for best production of crayfish, or even drained to a very shallow depth for growing plants in the rich silty muck which invariably builds up on the bottom.

The sides of commercial ponds are often fairly steeply sloped, making it difficult for herons and egrets to wade along and catch young fish and crayfish, but these slopes are prone to erosion in heavy rains so should be planted with grasses or more permanent cover immediately after construction. This won't stop more serious predators such as cormorants, which can only be kept out by netting over the pond (see discussion in chapter 5).

Deeper dams are used in commercial aquaculture as water reservoirs only, to fill or top-up culture ponds. Although it would be possible to raise a few more fish or crayfish in them, these are more difficult to harvest and also increase the chance of introducing diseases or parasites into the culture ponds downstream. As it is also difficult and expensive to cover over larger dams, a stocked water reserve on a fish farm is mostly a liability as it would mainly attract cormorants and other predatory pests.

## *Moving water*

The speed with which a pond can be filled is important, and even more important if you are emptying it for a drain harvest. The longer the crop is exposed to sunlight, predatory birds and changeable weather conditions, the more likely you are to have problems. For this reason, the ability to move water efficiently, and to a lesser degree inexpensively, is critical in any larger-scale operation.

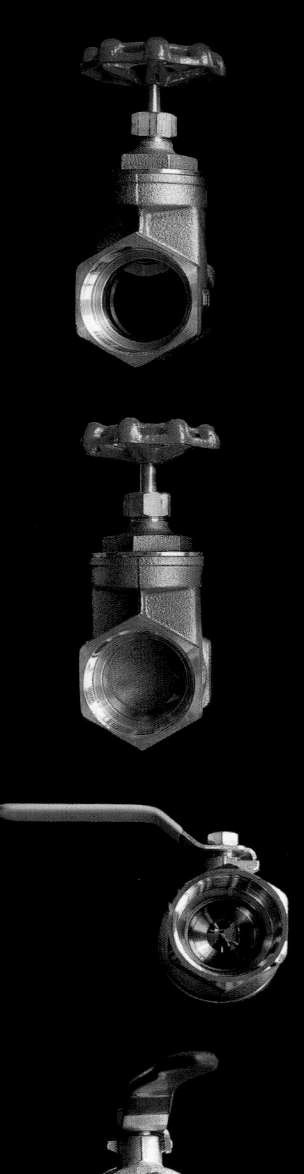

▲ *Flood valves with a tap-type handle on top, and gate valves with a swivelling handle, in both open and closed positions.*

▶ *A larger dam suitable for many forms of aquaculture, with a smaller dam in the foreground which is marginal for anything more than a few crayfish.*

▲ *Dam wall erosion caused by cattle has shortened the useful life of this dam by about two-thirds.*

The rate at which ponds with built-in drainage systems empty is dependent on pipe diameter, with larger pipes conveying proportionally greater flows. The faster the water flows through the pipe, the greater the effect of friction will be, which is why large diameter pipes are necessary for best flow rates – especially if using a pump.

Compared to a 50-millimetre diameter pipe, a 75-millimetre pipe carries more than twice the water, with only half as much again frictional contact with the inside of the pipe for the volume of water moving it. A 100-millimetre pipe carries four times the volume of water and has only half the frictional contact per unit volume. Any pipe less than 50 millimetres in diameter is of little use in all but the smallest ponds.

If you're only looking at moving around a few hundred to perhaps as much as a few thousand litres per hour, a 50-millimetre pipe is probably adequate. For any flow rate above 10 000 litres per hour, a diameter of 90 to 100 millimetres would be required, but for most commercial-sized ponds the minimum outlet diameter should be at least 150 millimetres.

Where portability is important, black polypipe comes in long lengths and can be coiled, but is only available up to 90 millimetres diameter. These larger diameters are relatively flimsy for the volume of water they can carry, so they shouldn't be built into dam walls where traffic passes overhead because they may collapse. The much more expensive and larger-diameter PVC pipes come as rigid units in shorter sections. While these aren't really practical for portability, they are strong enough to support vehicle traffic if buried deeply enough in a wall.

Pipes should not just be laid in a trench through a dam wall and then buried, because water will seep along their outside shell, gradually eroding a path through the wall. Even a piece of fencing wire running through a wall will do this with time, effectively creating a pipeline which will erode the wall once water starts gushing through it. To prevent such seepage, a series of baffles should be glued or clamped along a buried pipe. This is then placed into the trench, which must be filled and compacted by hand tools only. A crowbar will compact soil adequately around the pipe, but it delivers a mighty blow so must be used carefully around PVC.

The intake of any built-in drainage system must be screened in some way so that the crop doesn't just swim down the pipe with the water, and the size of the mesh used should stop even the smallest harvestable animals from escaping. Coarser, heavier screens placed before the finest ones will catch larger pieces of debris, and reduce potential damage to finer, weaker meshes. If water is brought into the pond through a pipe system from nearby natural waters where other (undesirable) fish species are present, it is usually necessary to mesh the intake there too.

The outlet of any drain holding water back is sometimes sealed by a plug, but these are fairly easily knocked loose if there is any pressure behind them, with disastrous results. Valves are a much safer way to seal pipes. Gate valves are less expensive but take time to crank open, and may seize up if not used for some time so considerable strength is then needed to open them. The more expensive ball valves use a longer shaft rotating 90° to open a hollow ball, and don't require any particular wrist strength to open.

Installing 'plumbing' takes time and is labour intensive, so ponds with no built-in drainage system are much less expensive to make. However, they are also less convenient to empty and on flatter sites must be pumped dry. Petrol-driven fire pumps are readily available, and are lightweight for their power so they can be used anywhere. Their inlet and outlet fittings are usually only around 40 millimetres in diameter, which limits the amount of water than can be moved through them, and as they are designed primarily to shift water at high pressure through fire hoses, fuel costs are relatively high even if they are run at a low idle.

There are other types of petrol pumps which are still reasonably portable but have larger fittings to take much larger pipes, for example axial transfer types which are designed to shift large volumes quickly rather than create much pressure. On a commercial scale, any property with a tractor will probably choose a pump that can be run directly off the tractor engine, combining high portability with considerable power.

On steeper sites a siphon can be used to drain ponds without fuel costs, and even a 50-millimetre pipe can be used to drain a one-megalitre dam over a couple of weeks, with a fall as little as one metre below the lowest part of the dam floor. Siphons are tricky to set up, especially those filled through a tap at their highest point, as all connections must be perfectly airtight, and dissolved gases coming out of solution as the water warms up in the pipeline can break the flow in summer. As siphons have limited application in aquaculture, they are not discussed in any detail here.

## *Modifying existing dams for aquaculture*

Dams which have been made for agricultural purposes, from watering livestock to irrigation, are not designed for aquaculture, but most can be used to raise at least a few fish or crayfish for domestic use. To optimise results and to make harvesting easier, there are a few modifications which can be done. Fencing off livestock is essential if water quality is to be kept high (although some crayfish may flourish in a shallow dam where cattle wade and excrete), and the fenced area can also be used to plant a surround of trees. Not only will these help improve water quality, but when they are tall enough they will also discourage predatory birds and many types of duck, which prefer open flight paths into and out of a dam.

If livestock have been using a dam for a long time, the erosion they cause will usually create a deep layer of silt covering most of the bottom. Few fishes will have much use for this smothering, high-nutrient muck, which also makes it easier for cormorants and pelicans to catch them in the relatively shallow, muck-free water. Crayfishes like it even less, as the silt clogs their gills, so they are confined entirely to the fringes of the dam.

If the silt is soft enough, most of it can be removed by a sludge pump or a dredge, and if firmer it can be dug out by an excavator. The mud removed is high in nutrients, and it makes an excellent foundation for a vegetable garden, or a mulch which is free of terrestrial weed seed. An excavator can also be used to cut planting shelves into steeper banks if you are interested in polyculture, with the heavily planted areas discouraging wading predators such as egrets and herons.

Other improvements that can be made include artificial underwater shelters created using logs, old terracotta pipes, and even old tyres if these are still legal to use in your area. There has been some suggestion that tyres may leach toxic materials such as cadmium and lead, but they have been used for many years on crayfish farms without problems and, as crayfish are far more sensitive to small amounts of these toxins than humans, any amounts leached must be miniscule indeed.

## *Reading further*

*Design and construction of small earth dams* by KD Nelson was originally published in 1985 (Inkata Press) and has been reprinted since, but *Farm dams: planning, construction and maintenance* by B Lewis (Landlinks Press, 2002) is more readily available at the present time. Either book is well worth reading if you plan earthworks on a large scale. For a detailed explanation of the setting up of continuous and primed siphons, see my earlier book *Farming in ponds and dams: an introduction to freshwater aquaculture* (Lothian Books, 1994).

*Plants can be a significant source of dissolved oxygen (seen here as streams of fine bubbles) in clear, well-carbonated waters, though air movement alone will provide enough for most uncrowded, open ponds.*

- Oxygen, carbon dioxide and temperature
- Acid, alkaline and buffering
- Hardness and salinity
- A sea of wastes
- Size of pond and water quality
- City water
- Reading further

# Water quality

2

Just as terrestrial plants and animals live and breathe or respire in air, so do aquatic species in water. However, aquatic animals also excrete their liquid – and not-so-liquid – waste products into water, where these dissolve and trigger various biological processes which all use oxygen. The dissolved wastes are also toxic, though in a planted or biologically mature pond they will quickly be changed into much less dangerous substances by bacterial and plant action.

The bacteria that detoxify the water also use oxygen and, as this is usually in limited supply in underwater habitats, even this beneficial process can reduce the number of underwater animals that can be stocked in a pond. As the dissolved gases are important for so many aspects of water quality, they need to be considered before anything else, though as will be discussed later many of the factors affecting water quality are interrelated.

## *Oxygen, carbon dioxide and temperature*

All animals need oxygen and excrete carbon dioxide as a waste product from its use. While oxygen makes up around 20 per cent of the air we breathe, there is far less dissolved in water, and even this greatly reduced amount decreases as water temperatures increase – the warmer the water, the less in the way of dissolved gases it can hold. For most fishes, other than those which are able to breathe air directly, oxygen levels markedly lower than one part per thousand are lethal.

The maximum concentration of oxygen that water can dissolve at any particular temperature is called the saturation level. In water at around 4°C, saturation is reached at less than 9 parts oxygen per thousand of water by volume; while at 20°C, saturation is about two-thirds of this, and saturation levels drop more rapidly still as temperatures climb further. These are *maximum* levels – in most still waters, especially in the deeper parts of a pond, oxygen is considerably less abundant because the gas must diffuse into the water from the air, and so oxygen levels are highest near the surface.

It is often assumed that plants produce a large part of the oxygen dissolved in water by breaking carbon dioxide up, as part of the process called photosynthesis. However, this mainly happens in shallower waters with abundant light and dissolved carbonates available, and most waters aren't clear enough for much active plant growth to happen below around a metre. Also, at night when there is no light to fuel photosynthesis, plants shift to a different metabolic cycle where they *use* oxygen and release carbon dioxide as a waste product.

In part, this explains why fully submerged plants aren't much used in aquaculture ponds. Firstly, the water is usually too deep and not clear enough for them to produce any significant volume of oxygen. Secondly, in warm waters at night, when oxygen levels are already dangerously low, large numbers of submerged plants can use up literally all of the oxygen, with lethal results for fish and other aquatic animals that can't breathe air directly.

Many biological processes in freshwaters are fuelled by oxygen, and this is used up rapidly when a lot of biological activity is going on. For this reason, the simplest and single most effective way to encourage growth and biological waste processing is to increase the levels of dissolved oxygen, preferably to somewhere near saturation, as discussed in the next chapter.

Dissolved carbon dioxide levels in water are often considerably higher than in air, especially in ponds where no plants are present, though it also enters the water dissolved in raindrops from the atmosphere. Apart from being essential for plant growth, its main and most dramatic effect is on the acidity of the water, as it binds chemically with water molecules to produce weak carbonic acid (see next section). Too much dissolved carbon dioxide also reduces the breathing capacity of aquatic animals, though this is rarely a problem in uncrowded waters, and never in aerated situations.

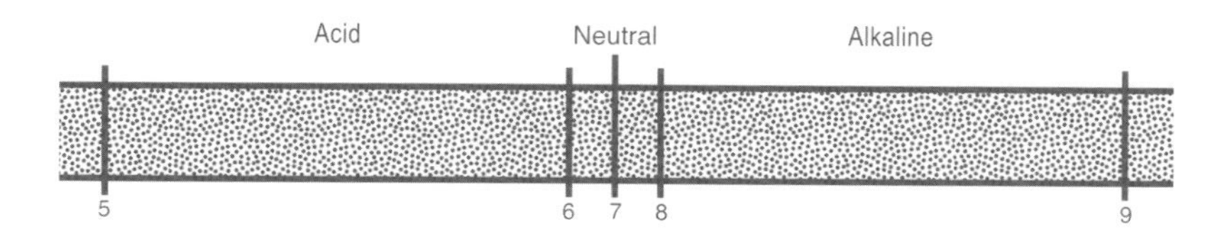


*The pH scale showing the exponential jumps between the numbers of the scale.*

Apart from biological factors, the main influence on levels of dissolved gases in ponds is stratification in warmer weather, where a hot layer of water forms on the surface of a still pond, with a considerably cooler layer below. The murkier the water, the more dramatic stratification is likely to be, because the heat of sunlight on a hot day is all concentrated in just the thin surface layer that it can penetrate.

However, even in clear waters a mildly warm stratified layer a metre deep is enough to reduce oxygen movement downwards. This is partly because the warmer surface layer holds much less oxygen than the cooler layer below could potentially absorb, but oxygen also moves less readily between the two layers of different temperatures. Mixing the layers through aeration is the usual cure for stratification, so that all levels remain at a fairly uniform temperature.

Even in a very large pond or a relatively small dam, stratification can be prevented from becoming a problem without any particular technology being required. Swimming (preferably freestyle) a few kilometres each afternoon on a hot day, as the sun's heat reduces, will not only mix oxygen into deeper layers, but is also the most appropriate and pleasant exercise for a hot day!

## *Acid, alkaline and buffering*

▲ *Aeration through air bubbles.*

▶ *Two edible plants (*Triglochin procera *and* Brasenia schreberi*) growing in mid-winter, as the water temperature near the bottom of their pond does not drop below 4°C. Water below this temperature begins to expand and rise until it forms a surface layer of ice.*

Pure water has virtually nothing dissolved in it, and is completely neutral – that is, it is neither acid nor alkaline. However, pure water doesn't occur in nature, and even raindrops falling to earth through the cleanest air pick up traces of carbon dioxide so they become, at least temporarily, slightly acid. Water running across or through the ground then picks up minerals, salts, and often soluble forms of organic matter.

These 'contaminants' are a perfectly normal and healthy part of water in most cases, if they aren't present in excessive amounts – for example very alkaline waters running off limestone, or city rainwater that has collected potent contaminants such as sulphur dioxide during its fall to earth. In fact, few farmed aquaculture animals will thrive in water which is too pure.

However, there are other extremes beyond which each species will not thrive, and for all practical purposes these can be measured and quantified in two ways. Measuring the total amount of salts and other minerals dissolved in water is discussed in the next section, while a simplified, practical account of measuring the strength of acid and alkaline waters is covered here.

The pH scale is used to measure degrees of acidity and alkalinity, starting at a benchmark of 7, which is perfectly neutral. The scale falling away to 6, 5, 4 etc. marks increasing acidity, while the scale

rising through 8, 9, 10 etc. measures increasing alkalinity. Although the scale looks to be a straight line, it is actually a logarithmic scale marking the concentration of free hydrogen ions, so each step away from neutral is ten times the previous step.

Thus, a pH of 5 is ten times as acid as pH 6, and a pH of 4 is a hundred times as acid as pH 6. Similarly, at the other extent of the scale, a pH of 9 is ten times as alkaline as pH 8, and a pH of 10 is a hundred times as alkaline as pH 8. In other words, measures on the far side of 6 and 8 are getting very rapidly into increasingly acid or alkaline waters, and in practical terms most aquatic animals and plants are best adapted to the 6 to 8 range – fairly close to neutral. They will also usually tolerate or adapt to small changes beyond this range if these happen slowly, though if water quality is allowed to drift beyond the ideal range, it can all too easily cross into the critical zone overnight.

Commercial scale operations with numerous ponds usually measure various water quality parameters including pH with complex and expensive electronic meters, requiring fairly regular recalibration using standardised buffer solutions. Such water testing devices may include a whole range of tests, from pH to hardness, conductivity, dissolved oxygen levels, etc., though all of these can also be bought as separate meters.

On a smaller scale, it is perfectly feasible to carry out pH tests with something as simple and inexpensive as a bromothymol blue kit, available at almost any pet shop or aquarium. This only covers a range

of 6 to 8, the preferred range for most aquaculture species including aquarium fishes, changing from green (neutral) to yellow at pH 6 or lower, and blue at ph 8 or higher, with a colour chart of graded colours giving fairly accurate readings for gradations of 0.2 between.

This will do as long as your water isn't always sitting at one or other extreme of this range, as bromothymol blue won't give you any way of telling if it is slipping below 6 or above 8. Universal indicator paper or liquid gives a much wider range of pH values, and though it is much less precise in gradations, it will at least give a clear indication if the water pH is slipping significantly beyond what is acceptable.

For small-scale aquaculture, you may not need to test the pH more than every year or two, unless conditions have obviously changed – for example, with reduced runoff during drought, or new upstream neighbours farming in new ways. Once you are familiar with the parameters of the runoff water entering your dam or pond, or your local tap water (which information should be available from your local water authority), these are likely to remain reasonably constant in uncrowded ponds where little supplementary feed is being added.

On a commercial scale, in most cases you would be crazy not to invest in a much more elaborate electronic test kit, preferably one that will also measure hardness and oxygen levels. This is because a relatively crowded pond has far more capacity for things to go wrong abruptly, than one just stocked to supply a family with fish. There are quite a few types of electronic meter available and no recommendations can be made here for any particular brand, but it is essential that all parts, calibration solutions and servicing should be readily available within your local area rather than interstate.

Water can become dangerously acidic if carbon dioxide levels build up too much, especially if it was acid to start with. To help prevent this problem, minerals such as coarsely crushed limestone, dolomite or coarse shellgrit can be added. Carbonic acid will react with the rocks and shell containing carbonates to form loosely bound bicarbonates which don't increase acidity, thus buffering the water against abrupt shifts of pH to the acid side.

A litre or two of coarsely crushed limestone scattered widely should be enough to buffer around 10 000 litres of soft water, without raising hardness to anything even resembling an undesirable level for most aquatic animals, or even plants. However, larger quantities may be needed for buffering where large numbers of animals are crowded into an intensive system and, as this will certainly add significantly to hardness levels, this concept will be discussed next.

## *Hardness and salinity*

Hardness is essentially a measure of the amount of dissolved mineral salts (mainly calcium and magnesium) present in water, while salinity is a measure of the sodium salts present. There is often a mix of both types present in most freshwaters, and the total concentration of these can usually be treated as a rough combined measure of hardness for all practical purposes.

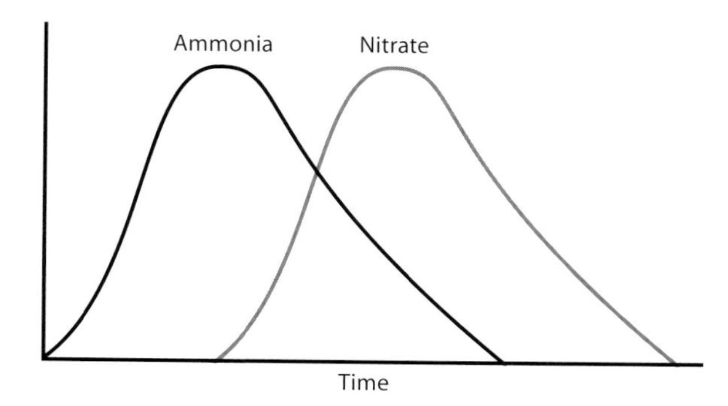


*The nitrogen cycle, showing increasing levels of ammonia until* Nitrosomonas *(which convert ammonia into nitrite) have built up in adequate numbers, triggering a similar cycle for* Nitrobacter *(which convert nitrite to relatively harmless nitrates). This cycle can take weeks or months to develop fully, with the fastest bacterial build-ups in warmer water.*

The fewer dissolved salts are present, the softer the water is said to be. This can be measured in various ways, though the easiest is to express hardness in terms of parts per million (ppm). Sea water averages a concentration approaching 35 000 ppm, the greater part of this in the form of sodium salts ( = salinity), while any water below around 200 ppm is pretty soft.

I doubt if any animals or plants of interest in aquaculture would be adversely affected by combined salinity and hardness figures up to 1000 ppm, and most would not be obviously set back even by 2000 ppm providing that this isn't all just alkaline mineral salts. Beyond this level, however, the growth of many species would be increasingly affected.

As in the case of dissolved gas and pH measurements, there are meters which can be used to measure hardness and salinity, whether separately or combined, but these are expensive and are unlikely to be used more than occasionally in a small-scale operation. Most of these measure conductivity rather than ppm, and as this does not convert precisely to ppm figures, it can be difficult to interpret. However, there are also simple aquarium test kits available which can be used to obtain approximate hardness figures in ppm, and these are all that most small-scale growers will need – you are unlikely to need to check this more than every year or two in an uncrowded pond.

Although explained here as separate concepts, the pH scale and measures of hardness or salinity are interrelated to some degree. Hard waters are often alkaline, while soft waters with a lot of biological activity going on in them are often on the acid side. Even saline waters are usually on the alkaline side, because it is rare to find sodium salts without at least some other types of mineral salt in most inland waters.

## *A sea of wastes*

Aquatic animals excrete in water, so they end up swimming in, and breathing, their own wastes. Much of their wastes are in the form of toxic ammonium and ammonia, and though ammonium is always more abundant this converts readily into the even more toxic variant ammonia as pH rises. Fortunately, many aquatic plants take up ammonium directly (see chapter 12), and even in an unplanted pond there are bacteria which deal with ammonium and its derived wastes through a process called nitrification (also referred to as the nitrogen cycle). Most discussions of this cycle simplify things by calling both forms ammonia, and this convention will be followed here except where the difference between the two is significant.

There are possibly many variations of this cycle using different bacteria, but in its simplest form ammonia is converted to nitrite by one group of bacteria, typically *Nitrosomonas* species. The nitrite itself isn't much less toxic than ammonia, and it also reduces the ability of fish blood to carry oxygen. As nitrite starts to build up in quantity, a second group of bacteria (typically *Nitrobacter*) increases in numbers, and converts the nitrite to much less toxic nitrates. Some fishes, such as trout, are more sensitive than others to nitrate, but it rarely causes problems even on the most crowded trout farm because ponds are usually being flushed at a considerable rate.

*▼ A backyard swimming pool modified for biological filtration, with small pumps mounted on upright pipes driving an undergravel filter which covers the entire floor.*

Nitrate can be used directly by aquatic plants and plant-like organisms including the single-celled algae which can turn water green, though these will use ammonia (in the form of ammonium) in preference if given a choice. This is because it takes energy on their part to reverse the cycle and convert nitrate back to ammonia once it has been taken up. Once the nitrogen cycle has established, toxic waste products are converted to nitrate rapidly, but as *Nitrobacter* are relatively sensitive to cold or excessive acidity (and nitrification lowers pH) the cycle can stall in adverse conditions, leading to a rapid build-up of toxic nitrites.

In a large pond stocked with young animals that have a lot of growing to do, establishing the nitrogen cycle is not usually a problem, as the small size of the starting stock means they won't produce enough waste to become dangerous before the nitrogen cycle establishes. As they grow, the number of bacteria already established can easily catch up with, and keep ahead of, the gradually increasing nitrogenous waste loads. It is only when large numbers of larger animals are shifted into a newly set-up pond that this is likely to become a danger. Such problems are also common in smaller, more crowded water bodies, particularly newly set-up aquaria which have been overstocked prematurely.

There is no treatment that will speed up the nitrogen cycle dramatically, and though there are bacterial products which are said to accelerate the final result I have seen no convincing evidence that these make much difference. Both *Nitrosomonas* and *Nitrobacter* are present in all types of freshwater (and seawater) naturally, and don't need to be introduced, though shifting a biologically matured filter from an established aquarium to a new one may shorten the maturity process by a few weeks.

Although nitrate is relatively non-toxic, it will build up gradually in a pond until it begins to affect plants and animals alike. It is also possible that other growth-retarding chemicals are produced, whether by the animals themselves or by the breakdown of other, as yet unstudied, waste products. Such shadowy but possibly real inhibitors have long been suspected to reduce growth of fish in smaller aquariums, and in plants an increasing number of studies are turning up allelopathic (also known as antagonistic) chemicals which some species produce to inhibit the growth of others.

Flushing is the simplest and crudest way to reduce waste product build-up in a pond, and while this can waste water, it can also be a source of combined irrigation and fertilisation for terrestrial plants. In a backyard situation where vegetables are also grown, a pond can be an asset in which water is first fertilised by fish wastes before being applied to the plants. In turn, the regular water changing allows greater crowding of schooling fishes, which (unlike battery hens) naturally congregate in larger groups.

In natural situations, bacteria carrying out nitrogen cycling attach themselves to any surface, and the greater the surface area for them to attach to, such as plants, roots and other finely divided objects, the more of them there will be. In a typical commercial pond with no plants, most of the bacteria will be associated with decomposing organic matter and detritus on the pond floor. Bacterial numbers can also be increased by providing artificial sites with numerous surfaces, for example various types of biological filter. Although effective, these are mainly only used in intensive aquaculture systems, which are discussed in the next chapter.

## *Size of pond and water quality*

Large ponds and small ones are very different in many ways, and the smaller the pond, the greater the expertise and attention of the owner must be to keep it in good health. For aquaculture purposes where any type of animal is to be raised, it is unlikely that anything smaller than around a thousand litres is practical, and even at that size there can be considerable and stressful changes in temperature between day and night in some climate zones.

Plant ponds can be smaller and shallower, to as little as 20 centimetres deep, but for animals other than ornamental fishes anything less than 5000 litres and a depth of 60 centimetres would need a filtration system or constant flushing to keep ahead of biological wastes. Feeding in such small ponds is also essential as larger fish would never be able to sustain themselves on natural foods in this situation, let alone grow.

The more you feed fish in a small pond, the more rapidly water quality will deteriorate, and aquaculture in small ponds is a balancing act between these two factors. If you allow water quality to drop too often or for too long, the growth of aquatic animals can be permanently affected. On the other hand, if you feed inadequately or infrequently, the animals will be much slower growing than in optimal conditions. There is no simple way to balance these factors, other than through constant and careful observation, with daily adjustment of feeding and water-changing needs.

Food quality also becomes more critical for small ponds, as commercial feeds deteriorate with time as will be discussed in chapter 4, so there is no point buying and storing large bags of these at a time – it would take many years for a small number of fish to use them up. Instead, you would be better off learning to make your own foods from fresh ingredients, and then freezing the surplus.

As with feeding and water quality, stocking rates for small ponds are a combination of guesswork and observation, but the same basic principles apply as for larger ponds and dams. Start with small numbers and work your way up gradually, allowing for the extra feeding (and its associated wastes) that will become necessary as the young animals grow. Aeration has already been mentioned as the simplest technological way of increasing the number of animals you can stock, and is particularly useful in crowded ponds during warm weather.

While you can produce some animals in smaller ponds by paying strict attention to these details, don't expect more than the occasional meal from them. Unfortunately, there is a widespread amateur belief that elaborate filtration and water recycling techniques make it possible to stock extremely heavily, regardless of species. In practice, most people who believe they can crowd fish in small ponds and still have good growth rates, generally don't bother keeping up the constant maintenance and surveillance required, so their aquacultures die out rapidly and they give up.

However, these notions are kept alive by various alternative technology books, the most overdone of which is a US edition of a self-sufficiency guide, which recommends stocking small ponds with 'about 1 pound of fingerlings for each cubic foot of water – more could overload the system [!]'. At this already critically high rate, if grown to an edible size, there would be about a kilogram of fish for every litre of water – though of course most fish would die long before reaching such a density.

## *City water*

Piped water in cities has invariably been treated in various ways to eliminate micro-organisms potentially dangerous to human health, and the chemicals often used can be dangerous to aquatic animals too. The most frequent problem chemical is chlorine, which evaporates as a gas just by allowing the water to stand for two or three days before using it for water changes, or before putting fishes or crayfish in it.

More dangerous is chloramine, a stabilised chlorine-containing compound which has apparently only been used experimentally in Australia to date, but if it is introduced more permanently it will be necessary to treat mains water before adding it to aquaculture ponds. There are various additives available to break down chloramine, particularly through aquarium stores, which are also a good source of information on what needs to be done to your local tap water to make it habitable for fishes. Your water authority will also be able to supply full details of what has gone into tap water, and recommend ways to remove anything that may be harmful to aquatic animals.

As many smaller ponds are backyard jobs in cities and suburbs, and their owners often use roof runoff water to keep them topped up, it is also worth considering rainwater quality here. All raindrops pick up at least some carbon dioxide as they fall, making them slightly but not dangerously acid. However, city air is often richer in carbon dioxide, and also the much more dangerous sulphur dioxide which creates more potent sulphuric acid when dissolved, so it is worth installing a simple shunt into rainwater downpipes to divert the first flows from a rainstorm along with their associated leaf litter and other rubbish. These are readily available from plumbing suppliers, as they are routinely used to keep detritus out of water tanks.

To make the most of rainwater flushing, the pond should be partly drained ahead of expected rain to allow 20–30 per cent refilling. If rainwater quality is high in your area, it may even be desirable to increase the change to at least 50 per cent, though it should be kept in mind that heavy rains will probably do this much flushing anyway, even without part-draining the pond.

## *Reading further*

Considerably more detailed discussions of various issues relating to water quality can be found in *Fish and invertebrate culture: water management in closed systems* (S Spotte, 2nd edition, published 1979 by Wiley Interscience) and also *Pond aquaculture water quality management* (CE Boyd & CS Turner, published 1998 by Kluwer Academic Publications, Boston). For a discussion of direct removal of ammonia by aquatic plants see *Ecology of the planted aquarium* (D Walstad, 2nd edition published 2003 by Echinodorus Publishing) which is available through ANGFA (see chapter 8).

The Australian and New Zealand guidelines for fresh and marine water quality suggest water quality standards for aquaculture (see <http://www.deh.gov.au/water/quality/nwqms/pubs/volume3-9-4.pdf>. These guidelines would be used by planning authorities, for instance, in setting conditions on a permit to discharge water from aquaculture ponds to a local waterway, but whether these are adequate is a moot point.

• Intensive aquaculture
• Aeration as the first step to increasing production
• Filters and recycling
• Water treatment wetlands
• Reading further

# Technology and water quality

# 3

*A small wetland for combined water treatment and habitat which absorbs most nutrients running off adjacent lawns.*

The simplest form of aquaculture is just dropping whatever animal you wish to raise into a pond or a dam, and harvesting a year or two later. This is known as extensive aquaculture, a rough-as-guts approach which does require some knowledge of the ecology and needs of the animal to succeed, though not much needs to be done apart from making sure that the water is already stocked with the preferred live foods for that species.

Yields from this primitive approach are not particularly high, and most growers try to increase the harvest through a combination of protection against predators, supplementary feeding, and stocking as heavily as is reasonable. The greater the number of animals being raised, the less likelihood there is that natural food supplies will be enough for adequate growth, so feeding must also be increased.

## *Intensive aquaculture*

Supplementary feeding adds to the amount of dissolved wastes in a pond, and as these have an effect on growth rates as well, one of the preoccupations of aquaculturists is using various forms of technology to increase the rate at which wastes are broken down or removed. It is possible to raise animals in crowded conditions by using diverse filter systems, and this high-tech approach is known as intensive aquaculture. Intensive systems are also referred to as recirculating systems, but this is misleading as they are very much dependent on an abundant source of clean water to keep their contents alive for more than a few days. Both terms are used interchangeably in this book.

▼ *A small trickle filter used to clean a bank of aquaria.*

Intensive aquaculture has sometimes been compared to the appalling conditions under which battery hens are raised, but there are significant differences. Many fishes school by nature, and for these it is not necessarily any hardship to be crowded together, as long as water quality is high. Even this isn't always essential, as some species such as jade perch are pre-adapted to crowded conditions in shrinking waterholes and droughts that may last years. On the other hand, some species may become aggressive when crowded, and no one tries keeping these intensively because it just doesn't work!

There are two major drawbacks to intensive systems, of which the less obvious is that no waste treatment system known will remove all traces of wastes from a crowded pond, so constant flushing with clean water is essential. Even with this, a certain 'background'

▲ *A paddlewheel aerator at work.*

level of toxic wastes will always be present, restricting the types of fishes or other animals to those which won't be stunted or disease-prone as a result of less-than-ideal water quality. The rate at which water must be replaced depends on the degree of crowding, but is usually at least 10 per cent per week.

The other major drawback of intensive aquaculture is cost, and the economics of such systems are discussed in chapter 7. Elaborate filtration systems are expensive, not only to construct, but also to run. Aquaculture trade journals often feature articles revealing new types of intensive system being offered as complete packages, and the promotional literature for these is (not surprisingly) always enthusiastic. This invariably includes a breakdown of initial cost and operating expenses, compared to projected yields, showing a steady and reliable return can be made from your investment.

However, a closer look at the maths behind these figures shows that annual returns are not great compared to the high cost of investment in *any* such system, and that a single crop failure can be enough to run the operator into the red permanently. As intensive systems are also vulnerable to disease, stress and water quality problems, any of which can take hold rapidly in crowded conditions, this form of high risk aquaculture is best left to people with considerable knowledge of the field, and who are able to be on hand virtually 24 hours a day.

In practice, most commercial aquaculture is semi-intensive, running on some combination of filtration, aeration and water-changing where the system used is far less likely to develop problems rapidly in the event of a power blackout, or malfunction of any critical component. The main exception here is trout farming, where constant flushing with cold, clean, oxygen-rich streamwater keeps water quality high without any need for technological intervention.

## *Aeration as the first step to increasing production*

The simplest method of increasing production in a pond is to aerate it, either by pumping air bubbles through the pond or by churning and otherwise circulating the water rapidly. This not only maximises the amount of oxygen dissolved, but also gets rid of excess carbon dioxide which can be harmful. The increased oxygen levels in turn allow heavier stocking rates and feeding, as the bacteria which convert toxic wastes into more innocuous substances are able to work faster. A further benefit is that animals which prefer or tolerate warmer waters will grow faster in warmer conditions, because their growth is not depressed by low oxygen levels at a time when they can eat and put on weight at a maximal rate.

There are various methods of getting oxygen into water, the most obvious of which is agitating or splashing the water's surface. Paddle aerators are still widely used for this purpose, mounted on floating rafts with their arms stirring the surface at a considerable rate, but these use a fair amount of electricity. Other, comparable aeration methods include pumping water to fall through a set of screens which break it up into spattering droplets, and small raft-mounted, anchored units with a churning propeller, rather like an outboard motor but running on electricity.

However, it is more economical in power terms to aerate the water simply by agitating, rather than lifting it above the surface, and this can be done by running an air stream through the water. There are several types of energy-efficient, high-volume but low-pressure air pumps which can be used for this purpose, though they will only work well in fairly shallow waters, with a functional depth limit of around 1.5 metres or less. At greater depths, more powerful pumps are needed, but these are increasingly expensive to run.

The air being pumped into the pond should be broken up into fairly fine bubbles by running it through some kind of porous hose or line such as Water Wik. A secured line of this kind can prevent stagnant patches from developing on the bottom, as the lifting bubbles generate an upward current which then spreads out at the surface. The smaller the surface area of the individual bubbles, the more rapidly they will release oxygen and absorb carbon dioxide on their upward journey, though a considerable amount of gas is also exchanged through the surface by the currents generated there.

Regardless of the type of aeration used, the aim is to deliver air as uniformly as possible to the whole pond, so that fish and crayfish can spread out rather than congregate where oxygen levels are higher. This in turn reduces aggression problems between animals forced into close proximity, and allows supplementary feeding to be uniformly spread as well, so that all animals have equal access to feed, not just the largest or most aggressive ones.

## *Filters and recycling*

The smaller a pond, or the more crowded it is, the greater the need to keep water quality high and break down toxic wastes as fast as possible. To accelerate these processes in intensive aquaculture, various combinations of filter system may be used, and the major types are described here.

### Mechanical filtration

Fine detritus and sediments can build up rapidly in a crowded pond, and much of this is from biological wastes such as droppings and tiny, uneaten food fragments. The sooner such particles are removed from the system the better, before they start to break down and decay, causing obvious water quality problems. A mechanical filter is simply any device which collects these solid wastes in some place where they can be removed easily. It is usually powered by a pump, arranged to generate currents that will waft detritus away from potentially stagnant corners of the pond and collect it in a container within the pond or pump it out to an adjacent filter.

It is essential to clean mechanical filters regularly by rinsing out or flushing the filter medium, ideally not long after feeding, so that the detritus collected does not have time to start decaying. Remember that these filters are like a vacuum cleaner only, and the decaying wastes in them will continue to release toxins back into the main pond or tank until they are physically removed.

Mechanical filters come in many shapes and sizes, and some use coarse sand or shell grit to collect sediments. For these heavier materials, it is important to be able to disconnect the filter and either rinse it or reverse the flow through it. This process, called backflushing, can also be combined with any necessary water changes needed. There are also many lighter, synthetic mats and felts which can be used to collect detritus, most of which can easily be removed from the filter and rinsed out for reuse many times over.

### Biological filtration

Once water has been pre-cleaned of larger particles that may clog finer filter media, the remaining dissolved wastes can be dealt with by bacteria, as described in the previous chapter, through biological filtration. As it takes time for bacteria to build up in numbers, usually weeks or sometimes even months, a newly established biological filter cannot be expected to handle too large a volume of wastes immediately. It can be conditioned by adding either a small number of animals to the pond initially, or a larger number of small animals which will grow along with the filter's capacity.

The more surface area there is for such bacteria to attach to, the more of them there will be, and the faster they can work on converting ammonia to (ultimately) nitrates. Even some of the mechanical filters already described will act to some degree as a biological filter, as long as some bacteria remain attached to the medium after it has been flushed of coarser detritus.

▲ *An elaborate mechanical filter incorporating an ultraviolet steriliser, suitable for a small aquaculture pond of up to 20 000 litres.*

One widely used type of in-pond or aquarium biological filter is the gravel or undergravel filter. In the undergravel version, a porous bottom plate or series of perforated pipes is covered with fine river gravel. This is connected to vertical pipes through which an upward current is forced either by water pumps or an airlift driven by an air pump. The pond water is drawn into the filter through the gravel where the bacteria are growing, through the bottom plate or pipes, and up through the lift pipes which return it to the pond. In gravel filters, the gravel is more usually kept in a separate section through which the water is run.

Maintenance of all gravel-type filters involves stirring up the gravel and siphoning off much of the debris that has built up. However, to remain functioning as a biological filter the gravel must not be cleaned out too frequently or you will lose the bacteria that are responsible for treating the water. The filter also acts to a degree as a mechanical filter as it collects fine particulate matter, and it is not desirable to remove this too efficiently either, because the detritus provides further attachment sites for the bacteria. On the other hand, the filter will only do a good job if most of the detritus is inert and won't break down into toxic stuff. To achieve this balance, feeding must be restricted.

▲ *Polypropylene bioblocks for use in a trickle filter, showing their large surface area.*

Another widely used biological filter is the trickle or wet/dry filter, which has long been used even for more heavy-duty waste water treatment such as sewage. Here, water from the pond is run into a separate filter compartment where coarser wastes are removed mechanically. The water then trickles through a perforated plate that spreads it evenly over the treatment chamber below, where the bacteria are established.

The trickle chamber is filled with any material that will support the bacteria, and has as much surface area as possible. These days, plastic biological filter media (for example, Bioballs) which are deeply divided to maximise the surface area, are made especially for such purposes. Their much-divided design means that a cubic metre stacked loosely has a surface area of around 300 square metres, and the wet, very well oxygenated surface area enables the breakdown of waste to proceed at a much greater rate than in even the most heavily aerated water.

To prevent water loss through evaporation, the trickle filter is usually sealed, with air being pumped in through the bottom. The trickling water is collected to be either pumped back directly to the pond or filtered again to catch any remaining sediments. Unlike most submerged filters, trickle filters will also actively break down nitrate.

Other variations on biological filters may involve using plant roots instead of plastic media, as these too are very finely divided and have a large surface area. These types either support the plants so their roots trail into a flow of water, or allow water to trickle down along them in a humid container. Plant roots are more than just an attachment site for bacteria, as many aquatic plants will also remove ammonium in their own right.

## Other filters

There are many other variations on biological filtration, some of which are high tech and high energy use, though these are designed to maximise attachment areas for bacteria in ways which are not possible with media that just lie around passively. For example, fluid-bed biofilters use fine sand which is kept in turbulent motion by powerful pumps. Typically, a small volume – say 50 litres of sand per cubic metre of water – is agitated to give a functional surface area of over 1000 square metres!

There are many other sorts of filters that deal with other types of wastes in different ways, and that can be used in conjunction with the more basic types already described. These include anaerobic filters, and protein skimmers which skim the oily surface film, though these don't work as well in fresh waters as in marine. Ion exchange, artificial resin and activated carbon filters don't work for long in crowded waters, and the high environmental costs of producing these media make them undesirable. Beyond filtration, some systems even pass water through an ultraviolet steriliser, though these are expensive to run and are no substitute for better pond hygiene.

Many of these more specialised filters and media are not necessarily cost-effective for the results they give. It is important to emphasise once again that not even the most elaborate combination of filter systems will ever be adequate to keep water quality high in an intensive system indefinitely. The most inexpensive way to augment the action of filters is by regularly replacing part of the water, ideally over a week or a fortnight, the nitrate-rich waste water being ideal for watering more earthbound crops.

## Water treatment wetlands

Another water quality issue which affects larger scale aquaculture systems that use considerable quantities of water is the quality of runoff downstream from the farm. This runoff, often rich in ammonia or nitrate, can be used as irrigation water for crops, but in some cases there is a legal requirement to return a certain volume of the water to wetlands and rivers downstream after restoring it to its original quality by removing wastes.

In these cases, running the water through an appropriately planted wetland is often all that is needed to restore water quality *providing* that it is large enough to deal with the volume of water passing through. This is partly done through the action of micro-organisms, and partly through luxury uptake of nutrients by plants (see chapter 12). Such wetlands can also be included as a part of recycling systems for intensively farmed ponds and dams, and may even make it possible to genuinely recycle water between ponds and wetlands if the wetlands are large enough and evaporation is not severe.

Many different designs of water treatment wetlands have been used worldwide, but most are variations on just a few basic themes. The most basic type of treatment wetland has water running through a heavily planted shallow basin or channel, and these are more than adequate for dealing with runoff from most aquaculture systems, as the water quality would be reasonable to start with. The flow should be as uniform as possible through all areas, and various combinations of baffles, channels and other dispersal devices may be used to keep it that way.

Water treatment wetlands must be precisely levelled, and in large-scale work laser grading is used for accuracy. On a smaller scale you can flood the area to be made into a wetland to a shallow depth once it has been roughly levelled. The water surface can then be used as the benchmark for levelling with a hand-held hoe. This ancient method is probably how large areas such as the bases of the Egyptian pyramids were made level. On a sloping site, a series of smaller, narrow channels zigzagging from one into the next downhill can act like a larger wetland, and the channels will reduce the need for baffles to maintain uniform flow.

Some wetlands use an open substrate such as coarse gravel to establish plants in, with water flowing beneath the surface of the gravel only. This is particularly useful in areas where mosquitoes are a problem, as it leaves no open water for them to breed in. The ultimate extension of sub-surface flow wetlands is to simply use the water to irrigate terrestrial plants, particularly tree plantations.

In another variation, water is run downwards through a loose substrate such as sand, which is densely packed with the living roots of reeds and rushes. The treated water is collected by pipes placed at the base of the root zone. These vertical-flow wetlands are similar in principle to trickle filters, and have mostly been used for small-scale water treatment as it is difficult to keep flow uniform over larger areas. As the root zone retains reasonable oxygen levels with this type of system, wastes are broken down more rapidly than in oxygen-poor environments.

Water may need pre-treatment (either by mechanical filtration or a settling/sedimentation pond) before being run through a water treatment wetland, especially if it carries appreciable quantities of sediment or other detritus. Yet no matter how good the pre-treatment, sediments and nutrients will build up gradually in every wetland, as will organic material from the plants themselves. All water treatment wetlands have a limited working life, and are likely to need at least partial scouring and replanting every few decades, depending on how quickly silt and decay materials build up.

## *Reading further*

Good accounts of aeration and filtration as applied to various closed systems can be found in *Aquaculture water reuse systems: engineering design and management* (edited by MB Timmons & TM Losordo, published 1994 by Elsevier Press) and less technical accounts in various editions of *The optimum aquarium* (K Horst and HE Kipper, original printing in English 1986 by Aquadocumenta).

For more detailed information on water treatment wetlands, some of the more useful recent works include *Constructed wetlands for wastewater in cold climates* (edited by U Mander & PD Jensen for Witpress, 2003), *Constructed wetlands for wastewater treatment in Europe* (edited by J Vymazal, H Brix, PF Cooper, MB Green & R Haberl for Backhuys Publishers, 1998), and for a broader Australian perspective see *Modern techniques in water and wastewater treatment* (edited by LO Kolarik and AJ Priestly for CSIRO, 1995).

- Live foods
- Prepared foods
- Making foods for small-scale use
- Reading further

*The essentials of high quality, homemade prepared foods for small-scale aquaculture: fresh fish, seaweed and freshwater algae, shrimps and free-range eggs to bind them.*

# Feeding

# 4

In the wild, fish stocks in any river or pool are restricted more by the amount of natural foods present than any other factor, as they are rarely able to build up to numbers that will significantly affect the available oxygen supply. In extensive aquaculture, stocking rates are chosen to be a reasonable match for the natural foods available in any pond or dam, after allowing around 50 per cent losses for ponds that aren't netted against predators.

The stocking rates suggested for the various species in chapters 10 and 11 are the approximate numbers that a certain area or volume of water is likely to be able to support without supplementary feeding, and to support larger numbers it will be necessary to feed the fish or crayfish appropriately. This chapter considers the many options available, from a small pond in the backyard, to commercial operations.

## *Live foods*

The optimal food for any species is the live (or occasionally scavenged) foods it would seek out in the wild, but providing these on a large scale is a daunting and often impossible task. Even for a small backyard pool, you would need many additional ponds to raise enough living prey to supply even a handful of moderately carnivorous fishes, and prepared foods will always be the mainstay of their diet.

On the other hand, some live foods in the form of plankton are likely to be available even in a fairly crowded pond. The nutrients present in these may be all that is needed to supplement prepared foods, which may not provide every micronutrient needed by a fish or crayfish. Commercial ponds are often drained between harvests and re-fertilised to encourage plankton blooms before any larger animals are added – this is looked at in more detail in chapter 9.

The best use of live foods is as a supplement to condition fish or crayfish for breeding, because the quality of their eggs and young, including hatchability and vigour, may depend on some poorly known or unstudied set of micronutrients in the preferred natural diet of that species. These foods may also offer other 'hidden' foods that are important – for example water fleas may contain essential nutrients in the form of the algae they feed on, and that a carnivorous fish also needs but is unable to obtain directly by browsing on tiny floating cells that it can't even see.

Thus, even the diet fed to the live food animals themselves may be significant in conditioning breeders. In the case of brine shrimp, it has been shown that those raised exclusively on yeast are neither as nutritious as those raised outdoors on a microalgal diet, nor are they eaten as eagerly by fish. The range of living things most native fishes will eat readily is huge, as many species are opportunistic feeders, and only a brief summary of some of the more readily available types is described here.

## Larger live foods

Water fleas (*Daphnia*, *Moina*, etc) are the most readily cultured freshwater food, and there are a number of species which will multiply quickly in an aquarium or small pond kept in reasonably constant conditions. They can be fed on microalgae for optimal nutritional value, supplemented if necessary with powdered yeast. The main trick with maintaining a constantly producing culture is to keep removing surplus animals for feeding, as too dense a population can result in dramatic drops in reproduction.

A steadily reproducing culture of water fleas is entirely female, with young females being born alive from their mothers while conditions remain good. However, stresses such as variable temperatures, overcrowding or falls in water level can trigger production of males, followed by breeding as the population starts to produce harder-shelled eggs that survive desiccation and other adverse conditions. Once this happens, the culture must be restarted, and if there are any warning signs ahead of time it is useful to have another all-female culture already underway for continuity of supply.

One disadvantage of this freshwater food is that water flea cultures may also develop freshwater predators, especially *Hydra*, a tiny, stinging relative of sea anemones. Although these are small enough that they can't harm even a fingerling, they are a very efficient predator on fish fry. The culture tank should be checked regularly for mats of *Hydra* clinging to the sides, and as one or two may be transferred with water fleas before they become conspicuous, these aren't an ideal live food to use in breeding ponds and tanks.

A crustacean that doesn't have the *Hydra* problem is the brine shrimp (*Artemia*), which is found and raised in saline inland waters. The drought-tolerant, cyst-like eggs of these species are imported in large numbers from the USA, and are an important aquaculture food. Brine shrimp will survive for some hours in a freshwater pond before being eaten, but their predators (should any accidentally be brought in with the eggs, which is extremely unlikely) will die before causing any harm.

The eggs are hatched in salt water at a temperature of around 20°C to 25°C, and are kept tumbling by a stream of air bubbles which also keep oxygen levels high. The tiny, newly hatched nauplii (an early larval stage of many crustaceans) can be fed direct to fish fry with large enough mouths, but their high salt content can kill the young of some species after a few days of feeding, so they are not ideal for all freshwater fishes. Fed on a diet of microalgae and yeast as for water fleas, brine shrimp grow rapidly and can be fed to the young of most larger native freshwater fishes in moderate amounts without problems.

Worms are an excellent conditioning food for many species of fish, and there are two main groups used. For smaller fishes, the aquatic black water worm is reliably available, and on a smaller scale can be cultured in a floating mat of the weedy, floating *Azolla* fern in a nutrient-rich pond

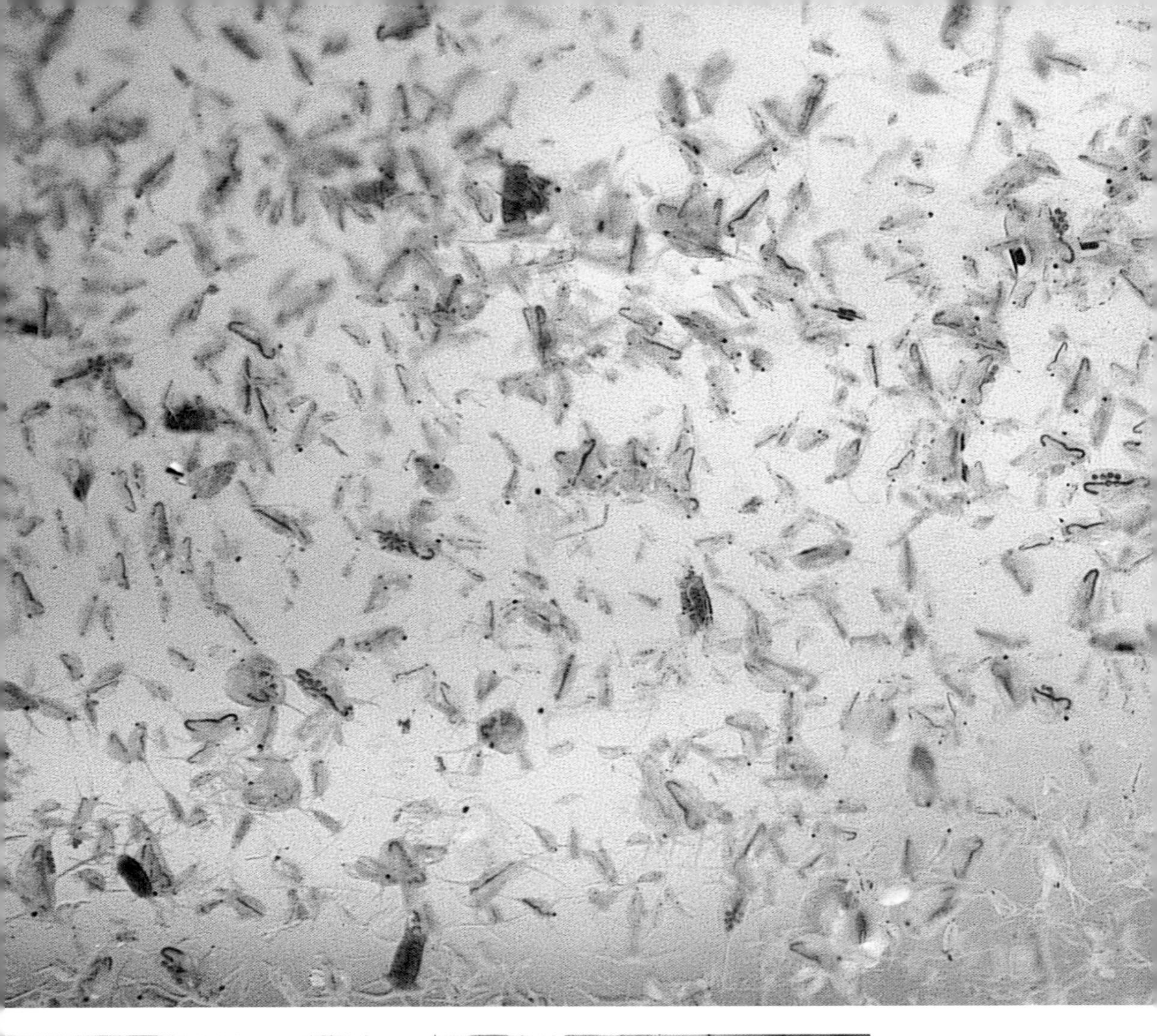

▲ *Copepods and waterfleas, two important planktonic live foods in fresh waters.*

◀ *Brine shrimp are raised in salt water, so they can be fed to freshwater fishes without any risk of transferring diseases or parasites.*

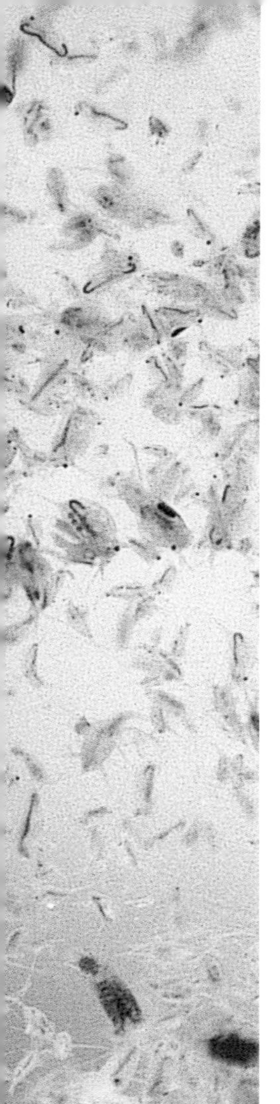

(but it should not be introduced onto larger dams as it can be difficult to get rid of). To harvest, fill a tall bucket with *Azolla*, cover with a blanket of fine sand, then fill the bucket with water until it is around 5 centimetres deep over the sand layer. The worms will crawl up through the sand when oxygen runs out below, and can be swirled around to form a tight-knit ball which can easily be lifted out. Those that aren't used immediately can be kept in a small amount of clean water in a large, sealed container in a fridge.

Earthworms are much larger, and can be raised in huge numbers in compost. There are many types of 'worm farm' available these days, including composting bins where worms do the work of turning and processing organic waste, and no recommendations for any specific type are made here. The most reliable type for raising as feed is the red worm, which is also the type most commonly offered as bait from fishing tackle suppliers.

Blue worms grow larger but are more erratic breeders, and there is no automatic advantage to producing a larger worm except as fishing bait, while tiger worms aren't eaten as readily by most aquaculture species. Earthworms should be purged of their gut contents before serving by keeping them in a clean medium such as damp sand overnight, and can also be chopped up and fed to smaller fishes.

## Smaller live foods

For smaller live foods and also for identifying other smaller organisms in a pond or dam, anyone with a serious interest in aquaculture should invest in at least one microscope. These don't need to be an expensive purchase, as good, basic models are readily available from China and Russia. A stereo microscope of anything from 12 to 20-times magnification is most useful for observing live foods, and it is fascinating to watch their three-dimensional movements in a shallow dish of water. A higher powered monocular microscope is also useful for smaller live foods, and is essential for identifying most disease organisms and parasites if you need to deal with these.

▶ *Stereo and monocular microscopes, with hand-lenses and a dissection kit – useful tools for identifying anything from live foods to disease organisms.*

Most smaller live foods are raised to feed newly hatched fish fry, and require good timing or continuously replaced cultures to keep up a steady supply. One of the easiest of these to culture, which is taken even by tiny fry, is the vinegar eel (*Turbatrix*). These are raised in a 50/50 mix of apple cider vinegar and water, with a few small, chopped pieces of apple added, and will build up into very large numbers within a couple of weeks. The culture will remain healthy for many months, and can be harvested by dangling a mop of acrylic wool into the vinegar, then rinsing this off in a small container of fresh water.

Microworms (*Pangrellus*) are comparably small and easily cultured, and I suspect they are even more nutritious as I have sometimes used these exclusively to raise the fry of some ornamental catfishes. They are cultured on cooked rolled oats or slices of bread soaked in milk, and kept in a sealed container as not everyone finds their sour smell pleasant. Once they are swarming up the sides, they can be dipped out with a fingertip, and a new culture should be started every two to three weeks.

For even smaller fry, you may need to culture microfoods such as single-celled organisms and rotifers, though this is usually only practical for young aquarium fishes. Cultures are difficult as you need fairly pure strains of desirable species to start with, and new cultures must be kept going at all times so the older ones can be replaced when contaminating organisms become too abundant. A simplified account of raising such specialised foods would be of little practical value, as each has specific needs and seasons, so readers with an interest in these will need to invest in at least one good plankton culture manual.

## Microfoods in pond culture

Microfoods are the ideal starter for newly hatched fish fry, but are difficult to culture on a very large scale. Instead, previously dried nursery ponds are filled a week or two before fry are due to hatch or be introduced into them, by which time they should have developed an excellent crop of tiny edible organisms. Draining and drying out the ponds also destroys most of the miniature predators of fish fry which build up otherwise, at least until the fry grow too large for them.

The dry nursery ponds are fertilised with hay and other organic matter (see chapter 9), or a crop of grass may be grown for this purpose if the ponds can be left dry for long enough. Once flooded, the vegetation soon begins to decay, providing an abundance of bacterial food for single-celled organisms and the more complex multi-celled rotifers soon after. As the fingerlings continue to grow, a bloom of larger creatures including water fleas and copepods usually follows. There is rarely any need to introduce any of these into the pond in the first place, as they appear spontaneously in the form of drought-resistant cysts blown in by the wind or hatched from the wetted pond floor.

By the time that micropredators, which could otherwise have made serious inroads into the numbers of fry, reappear the young fish will already be too large to be affected, so survival rates in nursery ponds are often high. By this stage the young fish may be large enough that there will now be insufficient live foods for them, and supplementary feeding with a fine, graded pelleted food suitable for their mouth size is often begun at this stage.

## *Prepared foods*

Most supplementary foods used in aquaculture are designed to keep well in dry form for weeks if not months. These must be not only a complete diet in their own right, but also palatable to the fish or crayfish being raised, though it may take some time to train them to recognise dry foods. Commercial aquacultures depend almost completely on prepared foods, of which there are now a wide range specifically researched to meet the needs of different grow-out industries.

Unlike live foods, which don't go off (but which can be killed if mistreated), prepared foods must be treated and used with care. The amount given must be checked carefully so that it is all eaten, and none remains to foul the water after. This amount must be seasonally adjusted, as most fishes eat less in cooler weather, increasing the feed as the fishes grow to keep up with their needs. To keep track of how much is being eaten, you can either use floating foods or, for species which feed only on the bottom, some kind of screen that can be raised to check what is left on it.

Learning how much to feed is a matter of close observation, although it will become almost intuitive with practice. In general, the ideal feeding regime is smaller amounts several times a day, but most aquatic animals are adaptable and will grow almost as well with a single larger feeding. In fact, if you are going away for a week or two it is better to leave aquatic animals unfed for that time than risk overfeeding if unexpected cold weather depresses their appetites.

Batches of most commercial feeds are produced regularly for freshness, as once they have been processed they begin to deteriorate slowly, even in the best conditions, and rapidly in hot, humid weather. This means that you should ideally only buy in enough feed for a few weeks at a time, possibly a few months if you can store them in cool, dry conditions. This may be a problem for smaller operators as feed is more expensive if bought and freighted in smaller amounts.

*Common examples of diverse living smaller foods (to different scales) which may appear spontaneously in recently refilled ponds; many of these can also be cultured.*

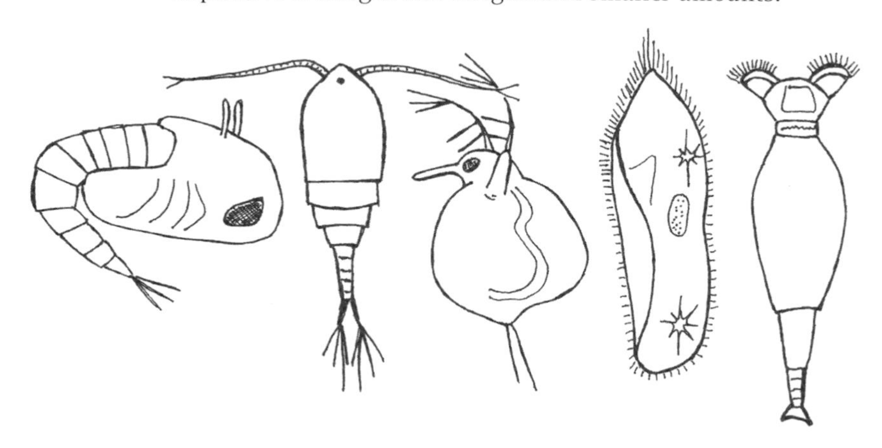

*Left to right: pupa of an anopheline mosquito, a copepod* (Microcyclops), *a water flea* (Moina), *a slipper animalcule* (Paramecium), *a rotifer* (Philodina).

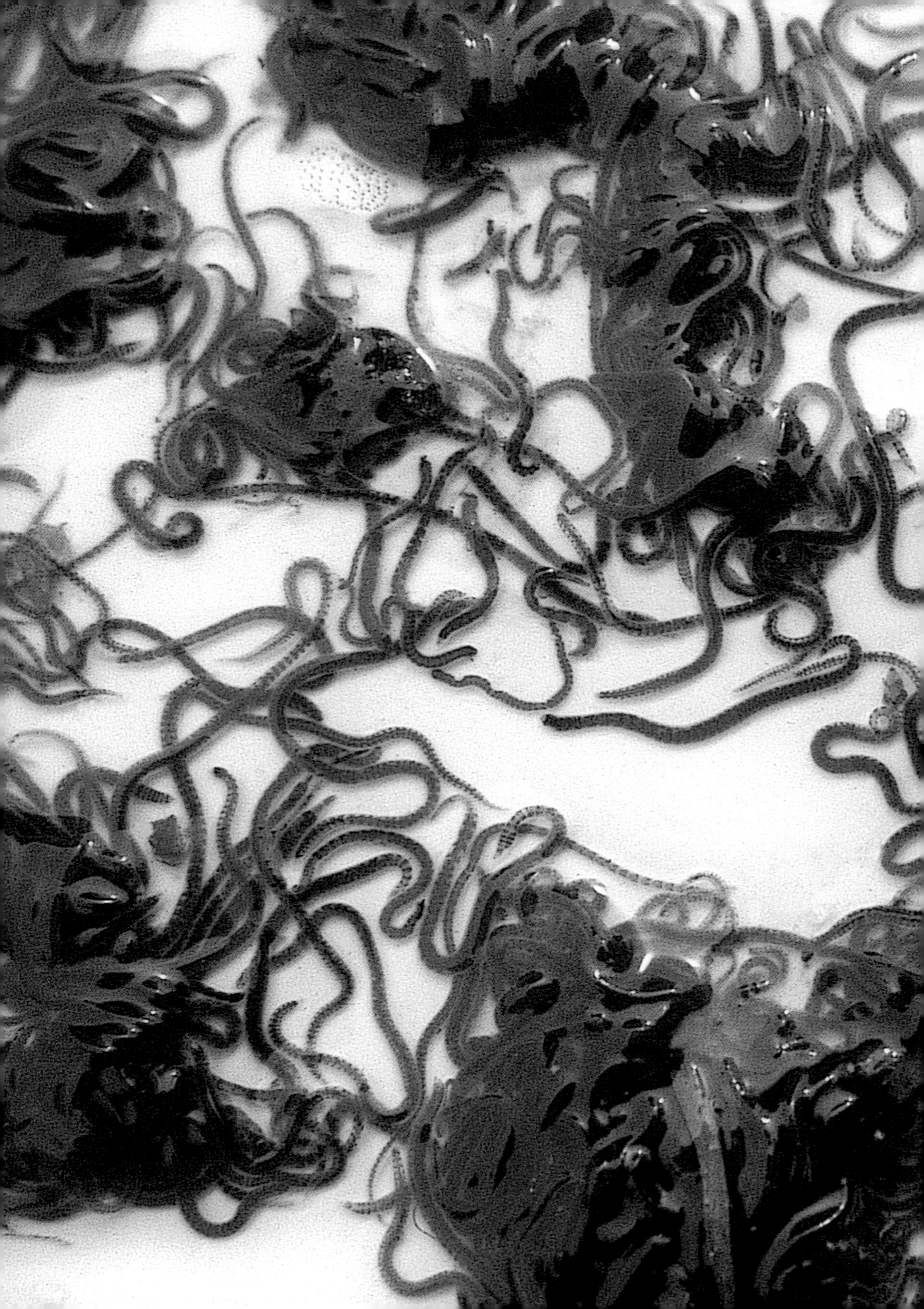

PREVIOUS PAGE *Black water worms are an aquatic species raised commercially in trout ponds.*

One possible solution if you are working on a reasonably large scale is to buy in larger and more economical amounts of freshly milled feed, then separate it into smaller sealed containers each holding a few weeks worth, and freeze these until they are needed. Deterioration is dramatically slowed in sub-zero temperatures, and even the occasional power failure for a day or two won't make any significant difference to quality unlike in the case of fresh foods, which may start to decay if thawed.

Feed is often the most expensive ongoing cost in commercial aquaculture, and it is important to make sure you are getting the best value from it. The feed you can obtain most readily locally is not necessarily the best for your operation, and it is well worth shopping around and testing several comparable products for palatability (that's for the animals to decide, not you!), conversion ratios, and related qualities.

Any supplier of bulk feed should be able to provide you with detailed analyses of each food they make, its ingredients, the species it is suitable for, and approximate conversion ratios (which indicate how much weight an animal puts on from eating a certain amount of the food, measured as a comparison of the weight of food eaten against the weight gain of the animal). The lower the conversion ratio, the more suitable the food is for the species you are raising. It also means you are not paying excessive freight for indigestible matter in the pellets, as a higher conversion ratio may mean some of the ingredients are just bulk and have no real food value.

By contemporary standards, a conversion ratio of 2:1 is not particularly impressive, a ratio down around 1.3:1 is fairly good, while a ratio of 1:1 is about as good as it gets. This just means that for each kilogram of *dry* food fed, you can raise one kilogram of animal. It doesn't suggest 100 per cent efficiency, because if you *also* dried the kilogram of fish it would only weigh around a tenth of the living animal. Even so, fish and crayfish are generally very efficient at converting feed into body weight compared to terrestrial animals because they don't need to keep their body at a constant temperature, or support their weight in water. This is one of the beauties of aquaculture.

## *Making foods for small-scale use*

The choice of commercial foods Australia-wide is now considerable, but it is not economical for a backyarder to buy a sack or two of high quality pellets and either have them go off long before they are finished, or occupy freezer space for months or even years. The other option is to make smaller quantities of your own feed, which does not have to be in pellet form, and can be frozen for use over a long period of time. The basis for most home-made foods is a binder, usually egg or sometimes gelatine, with various combinations of plant and animal matter finely processed and blended in depending on the species it is to be fed to. Spread the blended mix thinly on sheets of foil (without using oil), and then cook these at a very low temperature just adequate to set the ingredients. The finished product is then cut into meal-sized pieces and frozen.

Some of the more important ingredients for carnivorous species include entire, ungutted prawns or shrimps, which should be blended or chopped finely, and also fish, though the guts of these are better left out in most cases. I have often used freshwater fish and crayfish without problems in such foods, but these have been from my own ponds with no diseases or parasites present, and it would be safer to use marine equivalents to avoid the risk of introducing diseases or parasites from wild-caught freshwater organisms. As I have emphasised with hidden foods earlier, the quality of ingredients is important. Thus, eggs from free-range poultry will have a markedly better food value than those from battery hens, and wild-caught marine prawns will include a greater diversity of other foods in their digestive tracts than the notoriously poor-quality farmed ones.

Red meat is indigestible for most aquaculture animals, and if included can cause digestive problems and impacted digestive systems. This includes chicken and pork despite their being promoted as 'white meats', though their flesh is as red as any other if they are raised on free range – it is only the battery farming methods by which these are raised that makes them anaemic! For fishes which need or appreciate a vegetable component to their diet, the most important addition is algae. Freshwater algae such as mermaid's hair (*Spirogyra*) and similar green, stringy species are all fine, but avoid tough, dark-coloured and more amorphous blobs of green algae as many of these are actually blue-green algae (cyanobacteria), and some may be poisonous.

Marine algae, particularly kelp and green seaweeds, are also good in small amounts, adding a wide range of micronutrients. More readily available greens which will be appreciated by herbivorous and partly herbivorous fishes include lettuce, spinach, and even crushed peas if these are still reasonably sweet and tender.

## *Reading further*

Two good basic references for live foods are *Plankton culture manual* (by F Hoff and T Snell, original edition 1987), and *Identification manual for microalgae used in aquaculture*, both published by Florida Aqua Farms. Further accounts of more live foods, particularly smaller types used for feeding fry, can be found in my earlier book *Farming in ponds and dams*, and a much broader and more detailed overview of commercial feeding practices can be found in *Nutrition and feeding of fish* (T Lovell, 2nd edition, published 1998 by Kluwer Academic Publications).

Among the many good books on worm raising, *Earthworms in Australia* (D Murphy, published by Hyland House, 1993) is the one I have found most useful. There are a number of guides to using optical microscopes, though my original edition of *The optical microscope in biology* (S Bradbury, published by Edward Arnold in 1976) remains as useful as any for basic work.

*A rainbow trout showing multiple problems which probably contributed to its death even at a mature age, including deformity of the spine and an incidental bacterial infection of the dragging tail fin.*

# Permits and problems

As with raising any other type of animal or plant, whether for food or for profit, there is a wide variety of procedures and potential problems that take up time and effort in aquaculture. This is particularly true for commercial operations, where the difficulties may range from dealing with many levels of government, to increased disease potential and management requirements for relatively crowded ponds. This chapter covers a wide range of issues that don't fit readily into other sections of the book.

## *Observation and intuition*

Perhaps the subtlest problem in aquaculture is visibility – it is the only form of agriculture in which much of what goes on is hidden from human eyes by turbid water and reflection. Yet at a recent convention of fisheries biologists, it was commented that it is remarkable how much can be understood about what is happening below the surface without actually being able to see anything, using inference from a wide range of other types of clues.

In aquaculture, we must ideally acquire an almost intuitive perception of what goes on below the surface from those other clues, including changes in behaviour which may signify the first signs of disease, deteriorating water quality, or the even the onset of spawning. It is also important to make observation easier wherever possible, for example using floating foods to monitor changes in appetite, or using a retrievable screen to check for remaining food on the bottom as recommended in the last chapter.

Freshwater ponds in Australia are rarely particularly clear, and reflection from the surface is greater when you are lower down relative to the water's surface. This means that an observer walking up to the edge of the pond is unlikely to see anything but glare at first; on the other hand, fish can easily see the observer approaching from a distance because of refraction effects.

Refraction is the changing in direction of light as it passes through different clear media, and in the case of light passing from air to water the light is bent at a greater slant downwards. Fish don't get the reflection glare seen by the shore-bound observer, so they are able to see the observer approaching long before the observer can see them. Less

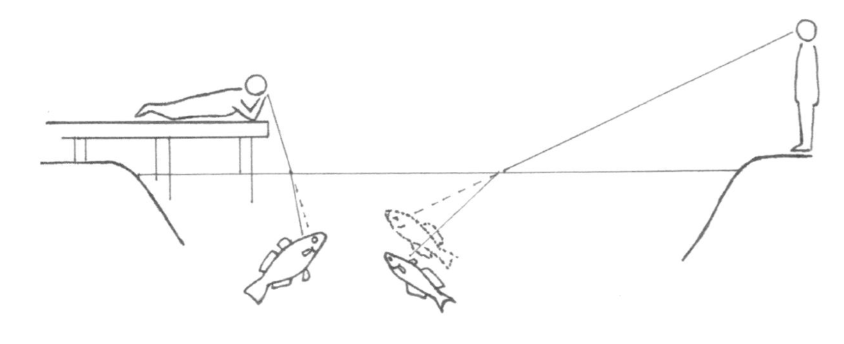

*The observer's view is subtly distorted as light passes from under water to the eye, and reflected glare at the shallower angle (right) also hides much of what could otherwise have been seen.*

importantly, the observer sees the fish higher up in the water column than it actually is, and anyone who has tried spearing fish from above the water's surface will know that they must aim lower down than the apparent position of the fish.

This is one of the reasons few aquaculturists actually wade or row around in ponds, as observations are more easily made from a higher perspective, and without disturbing the fish. A further reason is safety – waders easily fill with water and can act as an inverted parachute, towing the victim under. Neoprene waders which are skin-tight so they don't have this problem are readily available, but if I need to work in a pond I usually do so in a neoprene wetsuit longjohn with separate boots.

Both reflection and refraction can be minimised by a jetty over the pond, allowing the observer to use this as a type of hide, if necessary with observation ports through the floor so that aquatic animals don't realise they are being watched. Looking more-or-less down through the water's surface also reduces reflection, and what reflection remains can be almost eliminated with a pair of polarised sunglasses.

Aquarium observations are another excellent way of familiarising yourself with any aquatic animal you plan to raise, so you can come to understand its behaviour close up, and later be able to apply this more readily to the same species in a pond. Most aquaculture animals grow quite large, so it is generally best to start with younger ones in an aquarium, as a mature individual would need a very large tank to avoid stressing it. A general guide to keeping fishes and other animals in aquaria is included in chapter 8.

## *Permits*

Permits of one kind or another are required for most kinds of aquaculture, though you may be able to dispense with these if you are just stocking a dam with a few fish or crayfish for your own use. Over the past decade, the planning processes which must be gone through have been considerably reduced and simplified in most states, but regulations are still in a state of flux in many places so any attempt to summarise these would be outdated within a year or two.

There are two government levels which are likely to have input into any planning and permit processes for aquaculture. Your local council or shire will certainly require planning permits for commercial operations, and may also have controls on various other aspects of aquaculture from dam-making (see chapter 1) to species which can be introduced into your area.

For larger scale operations, most of the regulatory apparatus will be at the state or territory level, and in most cases these have now been streamlined so that all current information is available directly from the fisheries and/or game departments. It is essential to approach the state authorities before taking any other action towards a commercial

▲ *A simple clip tank used for observing smaller aquatic animals, with firetail gudgeons (*Hypseleotris galii*), a prolific fodder fish for much of eastern coastal Australia.*

▼ *A 5000-litre stock trough suitable for use as a fish nursery, quarantine tank or purging pond, with tropical waterlilies for water quality and reducing solar heat.*

▶ *Tilapia are widely cultured as a food fish in Asia, but are vermin in Australia and are illegal to possess in most states.*

▶ *(Bottom) Plague minnows – note the distinctive dark blue spot on the bellies of pregnant females, and the long fin below the body of the much smaller male near top centre.*

aquaculture farm, preferably even before buying land, as there may not be water quotas available for new developments in some areas.

Some common sense is necessary when approaching authorities, especially in terms of species to be farmed. Species which are already being farmed in the area and have established technologies behind them will usually present no difficulties as far as obtaining the necessary permits goes. However, if you are planning to experiment with species not yet being raised commercially, or which are subject to game laws (for example, river blackfish), the process may take much longer and become considerably more involved.

## *Diseases*

The more crowded a pond is, the greater the likelihood of disease and related problems, particularly those associated with less-than-optimal water conditions. Disease organisms which are specific to particular animals are discussed with those species in later chapters, while this section is an overview of the triggering factors for diseases, many of them related to poor water quality or crowding.

Many common and widespread diseases such as white spot and velvet disease are rarely seen in aquaculture ponds, because their free-swimming dispersal stage is tiny, and is probably preyed on by the diverse planktonic organisms found in outdoor ponds. In smaller environments such as aquaria, such diseases can run riot in a very short time, especially if triggered by a sudden fall in temperature of more than a few degrees. On the other hand, they are also easily recognised in the early stages, and treatment in a relatively small volume of water body is simple, using any of the wide range of specific medications available.

Bacterial and fungal diseases appearing in a larger pond are almost invariably a sign of poor water quality.

In a crowded pond, water conditions will never be completely perfect, and some growers are reported to have used medicated feeds to keep background problems from developing into full-blown infections. These 'cures' are antibiotics used inappropriately, an expensive and often ineffective approach which does not address the cause of the problem – poor hygiene. The excessive use of antibiotics in various areas of agriculture has also long been associated with overcrowding in suboptimal conditions and is, indirectly, a future risk to human health.

For this reason, it is better to tackle the source of the problem rather than try to mask developing symptoms. Bacterial infections are often associated with an overload of decay products from food, and sometimes an accumulation of solid wastes, which can only be dealt with by a significant water change combined with removal of as much organic waste matter as possible. In filtered ponds, bacterial infections suggest that the filter is being inadequately cleaned, or too infrequently, or that it just doesn't have the capacity to cope.

The water change may need to be repeated a week or two later, and badly affected stock may need to be disposed of as they will not necessarily recover even in optimal conditions, and are often not worth eating even if treatment is successful. A simple but permanent cure such as reducing stocking rates by just 20 per cent may be enough to stabilise the system at a healthier level. It is essential to adjust the feeding regime as well, whether this is just a matter of reducing feeding, or switching to some other food which is more fully utilised.

Before treating an apparent bacterial infection, make sure that the primary problem is the bacteria themselves, as some bacterial infections on fishes affected are incidental to a viral disease. To check this, you will either need to send a fresh specimen of the infected animals to the nearest veterinary lab specialising in fish, or learn to carry out preliminary identifications of disease problems yourself.

As many fish farms are a long way from the nearest appropriate specialist, and the combination of courier and laboratory charges can be high, anyone working on a smaller scale or just raising fish for their own table will be better off learning to identify and deal with the occasional problem themselves. While basic microscope and disease identification skills are also useful for growers working on a commercial scale, if there is any doubt in diagnosis the cost of identification by a professional laboratory is trivial compared to the cost of losing a whole crop.

Fungal problems are less common, and although there are some infectious fungal diseases, most aquatic fungi attack injuries rather than healthy fishes. Wounds can happen in the best regulated ponds, and can be treated by applying any of a number of fungicidal preparations to the wound. This should be done long before the fungus has had the chance to invade deeper body tissues, as the infection is often untreatable by that stage, and the infected fishes must be destroyed.

If fungal attacks become frequent in a pond, this is again a sign of overcrowding, and the subsequent aggression between individuals which can lead to secondary bacterial infections as well. However, even the most peaceable fish can accidentally damage each other with their spines if crowded, and an increase in such injuries is a sure sign that stocking rates need to be reduced.

Parasites are brought in either with the animal being farmed, or with one of its other hosts – particularly water snails, which can carry a wide range of transferable parasites. If you buy stock from an established fish farm, it is likely that any parasitic organisms they carry in the wild have been treated and eradicated already, so there is nothing more to transmit. In this case there should be no further problems if you can keep wild waterbirds away from the culture ponds, as some may carry larvae of certain parasitic organisms.

On the other hand, wild-collected stocks can potentially carry any of the full range of disease and parasitic organisms that affect their species, and must never be introduced directly into the future culture ponds. Instead, they should be maintained in observation or quarantine ponds for a long period of time, ideally as much as a year. Many parasitic problems will cure themselves, if there is no host animal (such as certain water snails) present for the next stage of their life cycle, but others can continue to multiply within the host indefinitely.

In the case of black spot disease, the parasite is the larvae of a freshwater mussel which eventually burrow out of the fishy host to become miniature mussels that can be disposed of long before they reach breeding size. If the mussels appear in a larger pond where it is impossible to locate them all, they should be destroyed while still small with a trace of copper sulfate, as drying out the pond would not worry most of them.

▶ *Netted trout ponds at Goulburn River Trout Farm with young fish feeding at the water's surface. This green netting is easy and inexpensive to set up and almost invisible, but must be removed for access to the pond.*

◀ *A partly buried tin barrier against water rats at Cultured Crustaceans.*

Quarantine ponds should be easily cleanable and sterilisable in the event of an intractable problem, and during their quarantine the wild stock should have any parasite and disease organisms present identified and dealt with. Once a quarantine pond is set up, and has been used as a way station for the introduction of stock into the main aquaculture ponds, it should also be used for any new stocks to be introduced even if these are from a reputable source, simply as a precautionary measure.

## *Predators and vermin*

### Introduced fishes

Most introduced freshwater fishes aren't particularly good to eat, and many are vermin which have had negative impacts on populations of native aquatic animals, and on entire ecosystems. The main species with any value in aquaculture is the rainbow trout (and to a much lesser degree, brown trout), though it would have been much better if these had been cultured only and never released into our rivers, where they threaten the survival of many smaller native fishes.

The most widespread of the introduced species is the plague minnow (*Gambusia holbrooki*), which has been introduced virtually worldwide as a supposed agent for mosquito control, even though it is not particularly effective at this. Wherever water is pumped from a stream into a pond, or fills dams with runoff from other water bodies upstream, there is a danger of introducing this unwelcome pest.

Plague minnows have been implicated in the disappearance of many smaller native fishes and frogs, partly because they are fast breeding, and partly because they are aggressive, nipping fins and eating fry of other species. A single pregnant female (they are livebearers) can generate a population of thousands over a long growing season, using much of the feed available in a pond and attacking any small fishes that are introduced.

Other, larger vermin including carp and (to a much lesser degree) tilapia are also widespread, and their young in particular are all too easily introduced to aquaculture ponds if these are filled with river water. Making sure that water being pumped or channelled into a pond is free of even the smallest unwanted fish fry is the best way to prevent unwanted visitors. In practice, this means running it through a screen as fine as mosquito netting and, as mesh this fine is easily clogged so it overflows, it should be checked regularly while water is moving through it.

▼ *Cormorants probably eat more cultured fishes and crayfish than all other predatory birds put together.*

Once introduced, the only guaranteed cure for most unwanted fishes is to drain the pond and dry it out, though some types of poison will often work if you can be absolutely certain that it has been mixed uniformly through the pond. Part-draining a pond or dam (if it can't be fully drained) will make it easier to distribute evenly any poison used, as well as reduce the quantity needed.

Rotenone has often been used in the past, as this breaks down into relatively harmless by-products within a few weeks, but is becoming increasingly difficult to find. Swimming pool chlorine and sodium hypochlorite should also kill most fish if applied to reach a concentration of around 4 parts per million (ppm), and the toxic chlorine component should evaporate into the air to reach a safe level for restocking within around a week.

## Indigenous fishes

Eels are the single most important predatory fish affecting aquaculture in the coastal regions of eastern to north-eastern Australia, mainly because they can travel overland in wet weather, and may appear in ponds even a kilometre or two from the nearest stream. They will eat anything from crayfish to small fish, and larger ones can even catch young ducklings.

It is extremely difficult to keep eels out of ponds in areas where they are abundant, though regular trapping with an opera-house net may keep their numbers low, and will also remove most of them before they become large enough to be a serious problem. In some situations, it may be better to take up eel farming rather than struggle eternally: remember that eels (especially short fins) are good eating!

## Other predators

Various other native animals will also feed in aquaculture ponds, and can make massive inroads into stocks. Many of these are protected, though a permit for trapping or shooting some species may be available in some circumstances. It is essential to check these options with your state game or fisheries authority before taking any steps towards removing any native animal which has become a problem.

Birds are the single worst problem in most situations, as they can fly in from considerable distances, and are intelligent enough to learn to avoid traps and shotguns. Cormorants are the worst of these, as they actively hunt underwater and can chase down huge numbers of crayfish, or even quite large fish. Muddy waters and tangles of plants or other shelters will make it harder for cormorants, though some casualties are still to be expected. Herons, storks and similar wading birds can be deterred to some extent by steep pond sides, or a dense planting in the shallows which interferes with their wading.

Some species of wild duck can also arrive in large numbers, and although they are inefficient predators, they may take large numbers of small crayfish. More worrying is the amount of undermining damage around the edges of a pond that dabbling ducks (especially the common black duck) can cause, and their impact on water quality through production of copious amounts of manure, most of it dumped into the water.

In an extensive aquaculture system or polyculture situation, stocking rates are relatively low, so the numbers of predators attracted will also usually tend to be low as there isn't enough prey to attract them on a regular basis. Nevertheless, casualties must be expected and allowed for, up to 80 per cent in some cases, by the time any animals introduced are ready to be harvested.

The only completely effective way to keep predatory birds at bay in any more intensive system is netting, which must be strained over the entire pond and also staked down along the sides, as herons and storks will push under loose netting. High-sided canopies make it possible to work under the net, but require heavier strainer posts and are expensive to set up. Where access is only required occasionally, a low net can be set up much more economically, but must be removed when access to the pond is needed.

Cormorants can often be discouraged by running tight lines at intervals of up to a metre (but preferably less) across a pond, as these poor flyers fly at a shallow angle and have trouble getting in and out between the lines. Scare lines are meant to be conspicuous and drive birds off as they vibrate in the breeze, but the fact that they must usually be spaced at 10-centimetre intervals suggests they are far from effective, and it is the canopy they create at this spacing which is more effective at keeping birds away.

Water rats are often a problem if the ponds are anywhere near a natural watercourse, and will travel overland some distance to feed if there is an abundant source of food to motivate them. Crayfish are a favoured prey of theirs, and they will eat large numbers in a sitting. As a native mammal, they are protected in most states, but permits can usually be obtained to catch them alive, and remove them elsewhere. Any traps used in water must stick out into the air as well, to prevent drowning of both water rats and also other animals which may be caught, from turtles to platypus.

Even trapping does little to reduce numbers where water rats are abundant, and it is better to choose a site at least a couple of kilometres away from the nearest wetland, through cleared country, rather than try to manage them. In the worst case, they can be excluded by a galvanised metal sheet reaching around half a metre above ground, and dug in around 30 centimetres, though this should be patrolled regularly to make sure a particularly determined rat hasn't burrowed under.

Freshwater turtles are another crayfish-fancying predator, also protected in most places as their numbers may be falling gradually but inexorably due to massive predation on their eggs by foxes. These are much more easily kept out of ponds by a fence, whether netting or

PREVIOUS PAGE
*Pelicans are an uncommon visitor to aquaculture ponds, but groups fishing co-operatively on large dams can capture an impressive number of large fish in a very short time.*

galvanised steel, angled slightly outwards to a height of half a metre so they can't climb it. Although they can dig when building nests, turtles are unlikely to burrow under.

Small fish and fry are vulnerable to a wide range of smaller predators, particularly insects such as backswimmers, and also mudeyes (the young of dragonflies) which can build up in considerable numbers in a pond which is never drained. When large numbers of young fish or crayfish are needed, it is usually necessary to keep these in a separate hatchery cum nursery area until they are large enough to introduce into unprotected ponds outdoors. These are discussed in the next chapter.

The most difficult predator to deal with is human – as aquaculture becomes more widespread, poachers with drag nets are becoming more common. The closer to a major city your ponds are, the more likely you are to be exposed to their activities at some time. While obvious deterrents such as fencing, guard dogs and alarms will help, some growers have used more subtle methods. For example, on one fish farm only women in skirts were allowed to feed the fish on the grounds that most poachers are probably male, while the male employees would throw rocks and were generally unpleasant to the fish whenever they were passing!

## *New pond syndrome*

Although water quality issues have already been dealt with in some detail, there is one which is best treated as a problem in its own right. This is 'new pond' syndrome, where a biologically immature pond blooms with a superabundant growth of algae. The worst of these are the cyanobacteria (commonly referred to as blue-green algae), many of which are toxic.

The cause of new pond syndrome is almost invariably overfertilisation, which is uncommon in most commercial aquaculture ponds where fingerlings are grown out. The problem is most common in planted ponds where an excess of fertiliser has been used, or the fertiliser has not been adequately buried so it won't leach into the water column. It is also seen where nutrient-rich runoff is washed into the pond from upstream, and can become a permanent problem in such situations.

To control (or better, prevent) new pond syndrome, all nutrients used to fertilise plants should be carefully incorporated into the planting medium so that they can't leach out. If nutrients are washing in from elsewhere, identifying the source of the problem is the first step to be taken, after which the offending runoff is either diverted, or a swale (effectively, a cut-off drain which remains dry except during times of rain) is used to absorb runoff so it doesn't reach the pond directly.

Barley straw is often recommended as a cure for excessive growth of algae in a pond, but this idea appears to have been based on a single, poorly designed study in the UK. Though the idea has been taken up with alacrity in some circles, I don't know of any further studies which actually substantiate the value of straw for this problem, and suspect that the natural maturing processes in a new pond have often caused the 'cure' for which the straw is credited. As an excessive amount of decaying straw can itself be a danger to water quality, it should be used sparingly if at all.

Other temporary cures which may keep algal growth under control while the pond stabilises include tadpoles of some species of frogs, and algae-eating fishes. There are increasing restrictions on moving tadpoles (and frogs) around, but if you have them in another pond on your property, particularly if there are too many for that pond, there is not generally a problem here. The only widely available native fish which eats algae is silver perch, and only the larger of these will eat it in any quantity.

## Inland saline waters

Many inland waters in Australia are markedly saline, and salinity is becoming an increasingly serious environmental problem across much of the country. To reduce salt intrusion into fresher waters needed for town and farm water supplies, an ever increasing number of groundwater interception schemes is being created in drier areas, particularly in Western Australia and western New South Wales. These expensive schemes divert saline groundwater to evaporation basins, where research is being done on aquaculture of marine and freshwater fishes, and of prawns which can be farmed successfully in these otherwise unproductive waters.

If these experiments can be converted into large-scale projects, there would be reduced pressure on coastal habitats, both aesthetically and in terms of pollution generated by surplus feeds and toxins such as from anti-fouling paints. There would also be obvious savings in sharing costs between aquaculture production and groundwater diversion.

Problems with inland saline waters fall into three areas. Firstly, inland temperatures fluctuate more than in areas nearer the coast, so any aquaculture venture must either stick to species which are tolerant of such variations, or use some kind of greenhouse effect such as plastic membranes to keep in warmth overnight.

Secondly, saline water holds much less oxygen than fresh, so aeration is likely to be essential no matter what species will be raised. One successful way of dealing with this has been in the use of floating raceways with airlifts driving water through a long holding cage where fish grow rapidly, but are more easily protected from predators and sunstroke. Floating pens are also being used to concentrate fish in a small area where they are more easily observed. As pollution from unused feed and wastes is a more pressing problem in relatively small, closed water bodies than in the sea, researchers are devising efficient ways of removing solid wastes which will hopefully later translate to marine cages with time.

For marine species, a major limiting factor has been the virtual absence of potassium in inland waters, which is essential for them to maintain a healthy osmotic balance – basically, keeping the right salts at the right levels in their tissues. Addition of agricultural potassium has proven to be adequate for such species as black bream and snapper, which also tolerate a surprisingly wide temperature range, but the range of edible fishes which will thrive in such ready-made 'seawater' is likely to be limited.

In Arizona, prawn farms are already producing a considerable tonnage of prawns from saline groundwater, and tests on black tiger prawns show some potential for these (and possibly other) species here. Although these are generally more cold-sensitive than the more rugged fish species, they can also be raised to a marketable size in a single season, with a possible option of raising rainbow trout in the same ponds over the colder months.

▶ *Mudeyes are a serious predator on young fish and fry, but take months to grow to a size where they can make substantial inroads.*

◀ *Backswimmers are active flyers and the larger ones can start killing small fish the moment they arrive in a pond.*

▲ *New pool syndrome in a natural swimming pool with floating mats of various algal species responding to excessive nutrient levels in the water.*

All further developments in evaporation dams are likely to be done on a commercial scale, but the present research also gives useful guidelines and pointers for smaller-scale growers with salinity problems of a greater or lesser degree. Seawater species are unlikely to thrive in smaller dams or ponds, which are in any case mostly much less saline than groundwater, and are unlikely to be practical for growers with a lesser salinity problem.

However, many inland aquatic species in Australia are pre-adapted to adverse conditions, and increasing salinity levels are common as drought dries out pools. For example, it has long been known that species such as yabbies and marron will not only thrive in salinities up to a third that of seawater (assuming adequate aeration), but also taste even better than their siblings from fresher waters. Similar results are reported for the native silver perch and introduced rainbow trout, and this is also likely to be true for a wide range of other species either being farmed already, or which may be farmed in the near future.

One area which has been neglected in inland saline research today is the prospect of using some suitable water bodies to produce brine shrimp (see chapter 4) on a large scale, either as an aquaculture food at their adult size, or to harvest the drought-resistant eggs to replace the ever-increasing numbers of these required for aquaculture, and presently being imported.

## Drought strategies

My original book on freshwater aquaculture came out in the middle of a serious drought that was affecting most of Australia, and one of the most frequently asked questions by interviewers was whether aquaculture was relevant at the time. Drought affects all forms of agriculture, and needs to be allowed for in similar ways, partly by reducing stocking to a sustainable level for the duration, and partly by consolidating all holdings into the best-favoured places.

Plants are readily built up again from seed stock, so they don't need much consideration. For example, a child's wading pool can be used to raise enough Chinese water chestnuts to restock a tenth of a hectare once the drought has broken. This tiny area will use a volume of water similar to that needed for an equal area of green lawn – but for a much better purpose.

Even fish and crayfish numbers can be rapidly rebuilt from a small holding stock of selected breeding specimens. Almost all farmed native species are already pre-adapted to some degree to the boom-and-bust cycle of drought, and are able to respond to improved conditions by breeding rapidly. Thus, a single pair of silver perch can produce as many as 300 000 fertile eggs in a spawning, more than enough to repopulate a substantial grow-out farm, while a mature female yabby may breed up to three times in a good season, producing over a thousand young.

Water is the all-important medium for all aquaculture operations, and unlike most agricultural resources it is fluid and fairly easily moved from place to place. This means that water resources can potentially be consolidated during a drought, by pumping or siphoning several part-empty dams or ponds into one full one. This is also an opportunity to bring in earthmoving equipment for removing silt and detritus build-ups from the empty dams.

The surface area of one full dam is considerably less than that of several half-full ones, and evaporation is therefore substantially reduced by consolidation. Evaporation can be reduced still further by using floating foam mats designed for this purpose, and mainly used in water supply reservoirs. However, such mats also cut off oxygen movement into the water, so they can only be used where aeration can be kept up at all times.

## *Alien introductions and conservation*

The aquaculture industry as a whole doesn't have a good track record on environmental issues. In inland waters many undesirable species have been released, whether accidentally or deliberately, permanently altering the ecology of most southern waters. A major component of the carp plague in so many south-eastern rivers trace their genetic lineage to escapees from an aquaculture strain brought into Victoria, while rainbow and brown trout were also bred and deliberately released for many decades, though admittedly for the 'benefit' of fishermen rather than for fish farmers.

Other introductions which have changed whole ecologies include yabbies in south-western Western Australia and also Kangaroo Island, where marron are also now abundant. In the tropics, various tilapias have bred up in huge numbers rapidly, and though most of these outbreaks have apparently been contained, there is still a potential for some of these species to create the kind of sterile near-monocultures now seen in many tropical countries. Potential weeds are discussed in chapter 12.

Where novel introductions outside the range of a species are being contemplated, it is essential to consider their potential impacts on the natural waterways around if they escape. It is no use pretending that they won't, because sooner or later a 100-year (or given the acceleration of climate change, perhaps even a 1000-year) flood will allow some individuals to get past even the most elaborate barriers. Further translocations into relatively pristine areas should be banned absolutely, and even in areas already well stocked with unwanted alien species, further introductions should be researched thoroughly.

The subtlest introductions haven't really been considered to date – new genes which have the potential to dilute those of wild populations once aquaculture animals escape into the outside world. As our knowledge of the genetics of wild populations increases, it is becoming obvious that even within a species there may be many very distinct geographic populations, all presumably better adapted to their regional conditions than their cousins from other areas. In some cases, these may even prove to be different species, and so introducing other populations into the same range is not only environmentally undesirable, but also genetically undesirable, diluting or altering gene pools that might otherwise be a useful future resource for aquaculture itself.

Yet even if aquaculture fish are from the same genetic population initially, with time they will change, as humans select from those which breed most readily or that grow most readily using prepared foods. In fish such as rainbow trout, which have been captive-bred for well over a century, the changes are glaringly obvious if fingerlings of the same age from wild and from cultured strains are kept in adjacent ponds. The cultured strain will grow faster on a prepared diet, and will come to the surface for food readily, unlike the smaller wild fish which will hide at the slightest disturbance.

Although native fishes have not been bred for so long that these effects are as pronounced, already some strains of fish such as silver perch are visibly stockier and heavier than their wild equivalents. With time, most cultured strains will become increasingly unlike their ancestors and increasingly unlikely to survive without artificial feeding, so that escapes or deliberate introductions of aquaculture strains will ultimately be incapable of thriving in the wild, let alone being capable of interbreeding with wild stocks.

## *Reading further*

Most books on fish diseases and parasites are far from user friendly, and many require considerable specialist knowledge to understand; however, *Fish disease: diagnosis and treatment* (EJ Noga, published 2000 by Iowa State University Press, Iowa) is reasonably easy to use. Two less technical books with a broad coverage, and using high quality colour photos for identification are *The manual of fish health* (C Andrews, A Exell & N Carrington, published 1988 by Salamander Books), and *Handbook of fish diseases* (D Untergasser, published 1989 by TFH Publications).

*Traps and nets including large box-type traps (two still folded), an opera-house trap, collapsible bait traps, a dip net, and a drop net with two rings.*

- Breeding
- Transporting and stocking aquatic animals
- Harvesting
- Cleaning and purging
- Reading further

# 6 From breeding to harvesting

In this chapter we look at both the beginning and the end points of sustainable freshwater aquaculture, from producing and introducing new stock to making the most of the harvest.

## *Breeding*

Few aquatic animals of interest to aquaculture will breed spontaneously in a dam or pond, and most growers find it more convenient to buy stock in from specialist breeders than to arrange special breeding ponds which may only be used for a short time each year. Crayfish are the main exception, and all three commercially grown species will restock their own ponds if an adequate number of breeding animals is left after each harvest.

Some native freshwater fishes such as silver perch are adapted to breeding after floods, and though the females may carry near-mature eggs for months, these won't ripen fully without the appropriate outside triggers. In the wild, such fishes would instinctively move upstream during flood periods, as their eggs will drift downstream and develop rapidly so the tiny fry can take advantage of the proliferation of microscopic food animals which will appear over the flooded plains around the river.

For species in which flooding triggers spawning, a combination of rising water levels and warm waters (generally above 22°C, or warmer for more tropical species) will do the trick, but this combination is difficult to arrange in ponds. If you have two separate ponds close together, flood conditions can be simulated by dropping the water level in the breeding pond a metre or more, and allowing it to warm up before 'flooding' it from the other pond by pumping or siphoning. To keep temperatures high, the incoming water must only be taken from the warm surface layer of the feeder pond.

For other fishes, the necessary conditions can be very different. For example, *Tandanus* catfish need relatively constant water levels (high evaporation rates and falling water may be enough to discourage them from spawning), and a relatively non-muddy bottom with pebbles for the male to build a raised nest from. Trout won't breed in dams, but can be manually fertilised once they carry ripe milt (sperm) and eggs by gently squeezing these out and mixing them together in clean, cold, strongly aerated water.

Some more difficult fishes (for example, those which carry near-ripe eggs for months while waiting for flood conditions) can be pushed into spawning by hormone injections, though these take some skill and should be done only by experts. These were originally prepared as an extract from the pituitary gland of carp, but more exact amounts of the same hormones that are used to increase fertility in humans are now used. Doses vary between fish species and must be prescribed by a vet. Often, only the female needs hormone treatment to trigger spawning, but males may also be given a smaller dose to 'tickle' them into action.

Regardless of species, the essential first step in breeding any aquatic animal is knowing its biology, and applying that knowledge to a specific breeding pond design. Although this is discussed where appropriate with the various animal species to be described later, it is important to emphasise that anyone planning to breed their own fingerlings for *any* fish needs to put considerable time and research into its requirements, in much more detail than any general account can give.

For the species of interest to aquaculture, spawning requires both male and female fish, the female ejecting ripe eggs while the male fertilises them by ejecting milt. The eggs are only receptive to sperm for a few minutes at the most, often less than that, and any which remain unfertilised will fungus over the next few days. As these can also infect fertilised eggs, any white or cloudy looking eggs should be removed once seen if they are being incubated in an aquarium or pond. If either of the parents is involved in looking after the eggs or raising the fry, these will usually remove unfertilised eggs without assistance.

Fish fry (and also young crayfish, if large numbers are needed) are often raised in nursery ponds to begin with, as these are more readily protected from predators. In the case of fry, even flying insects such as backswimmer bugs, or larvae such as dragonfly nymphs, can devour literally thousands before they grow too large to be taken. This means that nursery ponds must be netted in with mesh much finer than is needed to stop predatory birds. Once the young animals are large enough that insects won't bother them, they can be moved to outdoor ponds where only much coarser meshing against birds is needed.

## *Transporting and stocking aquatic animals*

Fish fingerlings and other smaller aquatic animals are usually transported in water in plastic bags, filled with just air for smaller numbers and shorter journeys, or oxygen for larger numbers on longer trips. As the fingerlings tend to pack themselves into corners where they may suffocate, or worse still puncture the bag with their spines, the corners are either taped up to round them off, or the bag is laid flat so the corners are out of water. Smaller, more expensive bags with built-in rounded corners are available from aquarium suppliers, but aren't large enough for the size and number of fish usually handled in commercial aquaculture.

The amount of water needed may vary from around a fifth of the bag space for a trip that won't take much above 24 hours, to half-full where a smaller number of fish are being sent on a much longer trip and it is essential to keep their wastes diluted to safe levels for the duration. To keep the fish relatively calm in darkness, and also to insulate against extremes of temperature (especially in air freight), the bags are packed into sealable foam boxes which must be kept out of sunlight at all times.

Many suppliers treat fingerlings with a bath against parasites, and also as a simple prophylactic, before packing them. A salt bath around 5000 ppm for an hour will do this adequately, and will also reduce stress

◀ *Hatchery at Goulburn River Trout Farm fully enclosed with fine mesh.*

▶ *Bagged silver perch ready for transportation – these have been double-bagged in case one bag develops a leak.*

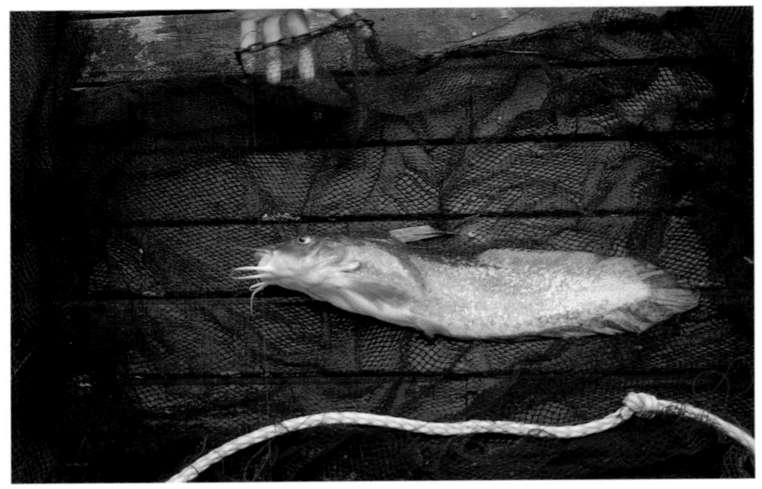

◀ *A seine (or drag) net showing the float line used to keep the upper edge floating, and a chain to weight the lower edge, with an eel-tailed catfish.*

▶ *Basic fishing tackle for harvesting dams – plastic fishing reels, diverse types of hooks, small leaden sinkers, plastic bubble floats and a yellow fish-hook disgorger.*

▼ *A fully netted pond at Cultivated Crustaceans keeps all predatory birds out.*

on the fish once they are bagged. However, some growers insist on using a very dilute solution of formalin, a much less environmentally friendly substance which can damage eyes and the nasal passages if breathed in while still at full strength.

On arrival, the water conditions in the bag must be gradually matched up with those in the pond or dam to be stocked. You can fairly safely assume that water conditions on the supplier's farm are close to optimal or they wouldn't still be in business. However, you will need to check both pH and hardness of your own waters to be certain there are no problems here, before introducing stock of any kind. Some suppliers recommend floating the fish bags for anything from 20 to 40 minutes to equalise temperatures, but I prefer to go further by adding from 20 to 30 per cent pond water to the volume in the bag, then repeating this half an hour later, and a third time a few minutes later.

Keep an eye out for obvious signs of distress during acclimatisation, then gradually tilt the bag so that the waters within mingle with those of the pond, and the fingerlings are able to swim out gradually – just tipping them out in one go only adds to stress as they enter their new home. Quarantine is not necessary if you are stocking a dam for the first time, but should be considered if other fishes are already established. However, having adult fishes already entrenched can be a problem if they are predators, and it is usually necessary to remove all of these beforehand unless you want to provide them with an expensive meal of live foods.

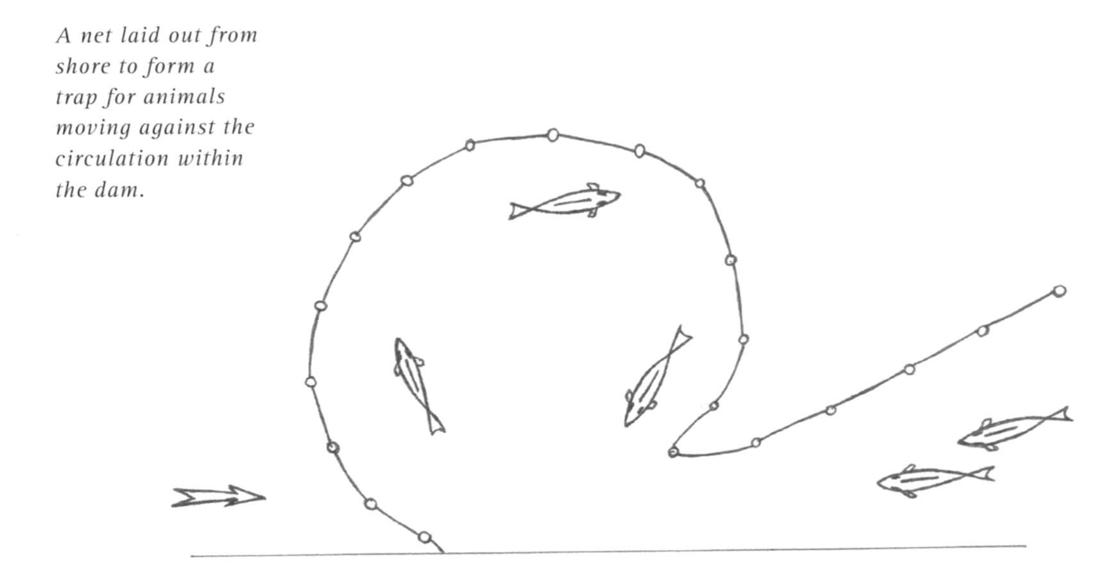

*A net laid out from shore to form a trap for animals moving against the circulation within the dam.*

## *Harvesting*

Harvesting the waters covers a disparate range of techniques. These can be as simple and pleasurable an activity as using a fishing line and baited hook to bring in what you need for the day. At the other extreme, the full-scale drain harvests of commercial aquaculture may need a small army to transfer the crop and process it as fast as possible.

### Fishing with a line

You don't have to be particularly skilled to use a fishing line effectively, nor do you need much in the way of equipment unless you fancy yourself as the sporting type. The most basic fishing set-up for freshwater ponds is a circular hand reel, with fishing line between 3 and 5 kilograms breaking strain – anything stronger is also coarser and more obvious to fish, as well as more difficult to tie to smaller hooks.

Hook sizes vary according to the type of fish you are planning to catch, and a good idea of the appropriate size range for any species can be found by reading any basic freshwater fishing book. However, there is an element of commonsense involved here and you can also get a good idea of hook size just by looking at the size of the fish's mouth.

If you are fishing for a species that feeds on the bottom, a small lead sinker is all that is needed to sink the bait, and this is slipped onto the line before tying the hook at the end. When the bait is taken, the line slides easily through the sinker, without dragging and alerting the fish to possible danger. For fish that feed from the surface, a bubble float takes the place of a sinker, and can be slid along the line so the bait can be raised or lowered to any level where fish are feeding. This type of float can also be squeezed to let water in, adding a bit of weight so the bait can be cast out some distance if that is where the fish are feeding.

Attaching the hook is the greatest skill you will need, as most random knots will weaken the line so that it breaks under pressure. Every basic fishing guide includes a range of knots suitable for nylon line, some of which are easily learned and memorised. Now all that remains is to firmly attach the chosen bait to the hook, and this should be one of the preferred foods of the fish you are trying to catch. Shrimps and small crayfish are readily taken by most aquaculture species, as are worms.

If you want to be particularly unsporting, you can train the fish to feed regularly where you will be fishing for them, by throwing in a few of whatever will be used as bait each day, at the time of day when you plan to fish. Alternatively, if you have been feeding them regularly for months, leave them hungry for a day or two before you go a-fishing.

Most pond-raised fish are as thick as a brick and will take the baited hook readily, so there is no great skill involved in sinking the hook to hold them, and you don't need to jerk the line hard or pull hard to land them. However, if the fish being caught are heavy, it may be useful to have a landing net handy to lift them out of the water, rather than put too much strain on the line itself. Hooks can be used even when the fish

▲ *A holding net used to purge a small catch of yabbies – the top must be enclosed as yabbies can climb mesh.*

are to be kept alive for some time before eating, by cutting the line and leaving the fish to disgorge the hook at its leisure. This may seem cruel, but most native freshwater fishes are carnivores which feed on a range of spiny animals including shrimps and crayfish, and inevitably some of these will leave spines stuck somewhere in the gut.

Fish must be able to eject such spines, just as most splinters will be ejected from skin in their own time, though the process seems faster in the case of fish. Barbless hooks are particularly easy to disgorge, but can only be used if the captures are pulled in immediately they take the bait, as they may fall out spontaneously if the line is too slack.

## Nets and traps

Netting a larger pond or a dam requires a net which reaches from one side to the other, with floats at the top to keep it on the surface, and weights on the lower edge to keep it on the bottom. If the dam is already low, for example during a drought, it may only take two people at either end of a net of less than 50 metres to pull it from one side to the other, collecting everything in its path, though a second pass may be needed to pick up any strays which got away the first time. For obvious reasons, fish such as trout, which can jump very effectively, are difficult to catch with a drag net.

In larger ponds with a considerable crop, the drag of the net will be far greater than even the strongest humans can reasonably handle, and in such cases a tractor on either bank will be required. Both machines must move slowly and evenly, or they are likely to lift the weights off the bottom in jerks, allowing most animals to escape. The disadvantage of netting large numbers of fish or crayfish in a single swoop with a net, is that the massed animals in the net as it is pulled ashore are far more likely to be damaged, reducing their saleability. Processing also becomes more urgent, as the entire crop is exposed to the air simultaneously.

Where netting an entire dam is impractical, such as in a large and deep water reservoir, or a polyculture pond with diverse plants along the shoreline, a net can be arranged to form a type of simple trap. Basically, it is run out from shore for a few metres, then parallel to the shore, with some kind of constriction to form a holding pen before the open end splays out again to divert passing fishes into the holding area.

All dams and larger ponds have a slow current circulating around them, which can be detected by dropping in a cork on a perfectly still day. Most native fishes and many crayfish instinctively move upstream against currents, however weak these may be, so a net trap like this is most effective if the open end faces 'downstream'. When the trap is to be closed, a rope attached to the open end is used to pull this ashore, and the entire net can then be pulled to shore in sections, like a drag net.

Other nets such as dip nets are less useful for large and fast-moving fishes, although they can be used effectively for catching small fishes hiding at the edge of masses of vegetation, and for catching crayfish off the ends of baited lines. Crayfish are also easily caught in drop nets, with two metal rings attached within, so that the baited net lies flat on the bottom. When the outer ring is pulled up to surround any crayfish within, the smaller ring drops below to form a deep pocket which is not easily escaped.

Traps are an easy way to catch and hold crayfish, and you don't need to be on hand at all times to use them. These all use the principle of an entrance such as a funnel or a vee narrowing as it enters a larger holding chamber. Such traps are usually baited with fresh meat (old meat does not work), which attracts most types of crayfish – even those which are

▼ *A holding cage used for purging fishes, usually attached under the jetty for shade.*

primarily vegetarian. There are increasing restrictions on the type and number of traps which can be used in public waters, but these rarely affect their usage on privately owned and stocked dams.

The most commonly used traps for crayfish in eastern Australia are the aptly named, upward-folding opera-house types. These will also catch some fishes such as eels and even *Tandanus*, and where turtles are present their tops should be allowed to stick up slightly above the water's surface, to allow the turtles to surface and breathe. In Western Australia, a larger, folding, box-type trap with a vee-shaped entrance along the top has proven to be useful for holding large numbers of yabbies. However, I haven't found these to be much more effective than opera-house traps and bait traps when smaller crops are being harvested.

Small, collapsible bait traps are used to catch smaller fishes as well as crayfish, and although larger versions aren't available commercially they can be made out of stiff mesh for larger fish. Some such as eels will push their way into such traps out of curiosity, or to take shelter if the body of the trap is made from a broad pipe or drum, with a funnel of wire mesh leading in from either end. Others will be attracted by certain baits, for example *Tandanus* catfish to crayfish tails.

### Drain harvests

Traps may be an effective way to catch a small number of fish or crayfish for home use, spreading the harvest over a long time, but in commercial aquaculture the aim is often to get everything out of a pond, processed, and sold on in as short a time as possible. For this reason, most commercial ponds are designed with a relatively flat bottom sloping gradually to a deeper sump that can either be pumped out or that has a drain already built in.

A large pond of this kind may take a day or two to drain, and this time should be calculated by working out drainage rate against pond volume, so that there are people present at all times during the final stages of emptying the pond. Left unattended, much of the crop could be gone by the time the pond is almost empty, especially if large flocks of cormorants live in the area.

It only takes one person to keep flying predators at bay, but dozens may be needed for the efficient handling of the final stages of harvesting the crop – grading crayfish by size and quality, gutting and processing fish rapidly, or loading live fish into a tanker with minimal delay, depending on what is to be done with the crop.

## *Cleaning and purging*

Freshly caught fishes can either be kept alive until needed, or gutted and either cooked or chilled as quickly as possible while their flavour is at its absolute best. Cooked and eaten immediately, even less popular fishes such as mullet can be excellent eating, yet left ungutted for a few hours longer they are not even worth keeping for bait. However, some fishes caught from farm dams have a distinctive off-flavour which is possibly the result of eating too much earthy material, or perhaps even some mild fungal infection from poor quality waters.

It is essential to get rid of this taste before sale or consumption, and most freshwater aquaculture crops benefit from purging for a few days before they are eaten, usually by being kept alive but unfed in clean water. For example, golden perch often develop a musty taste when raised in farm dams, which makes them much less palatable, but this will disappear after a week or two in a completely clean pond.

Freshwater crayfish especially benefit from purging, emptying their guts completely within a couple of days. Apart from some people having an aversion to the taste of the gut contents (which become stronger flavoured in animals which have been cooked hours before being eaten), they also look cleaner inside when cooked. As crayfish will live perfectly happily out of water if kept wet, purging for these can be done in prawn trays or similar containers with water dripping down on them, and as they aren't inclined to fight in these conditions they are less likely to damage each other than when purged in water.

The simplest way of purging fishes on a small scale is in a holding cage, which can be made of any kind of stiff mesh, suspended in a shadowed place such as under a jetty. Such cages should be large enough for even a fairly large fish to swim around in, and may need a cover to keep jumping species in. For animals which can climb, especially crayfish, a live bait net which closes tightly at the top is essential.

Purging fish on a larger scale is done in dedicated ponds, often with a recycling system to keep water quality high. The tank should be kept fairly dark as this is calming, and a cover is needed for jumping species. The process takes anything from one to two weeks, though this may need to be still longer at colder times of year.

Where adult fish are to be transported alive to markets, this is usually done in trucks with specially made live wells including provision for aeration. The aim with these is to get them into restaurants, markets and ultimately stomachs as quickly as possible, so their long-term wellbeing is not usually considered, but purging remains essential as the fewer waste products they produce in transit, the less stressed they will be.

## *Reading further*

For a more detailed discussion of insect predators on fish fry and fingerlings, see my earlier *Farming in ponds and dams*. For readers planning to use traps for harvesting their ponds, *Fish catching methods of the world* (my copy is the third edition, written by A von Brandt and published by Fishing News Books in 1984) contains a wealth of traditional designs which can be adapted to Australian conditions.

*Grow-out ponds at Goulburn River Trout Farm with trout being fed at near-freezing temperatures.*

- An overview of the industry
- Types of commercial aquaculture
- Other markets
- Intensive or extensive?
- Finance and investment
- Reading further

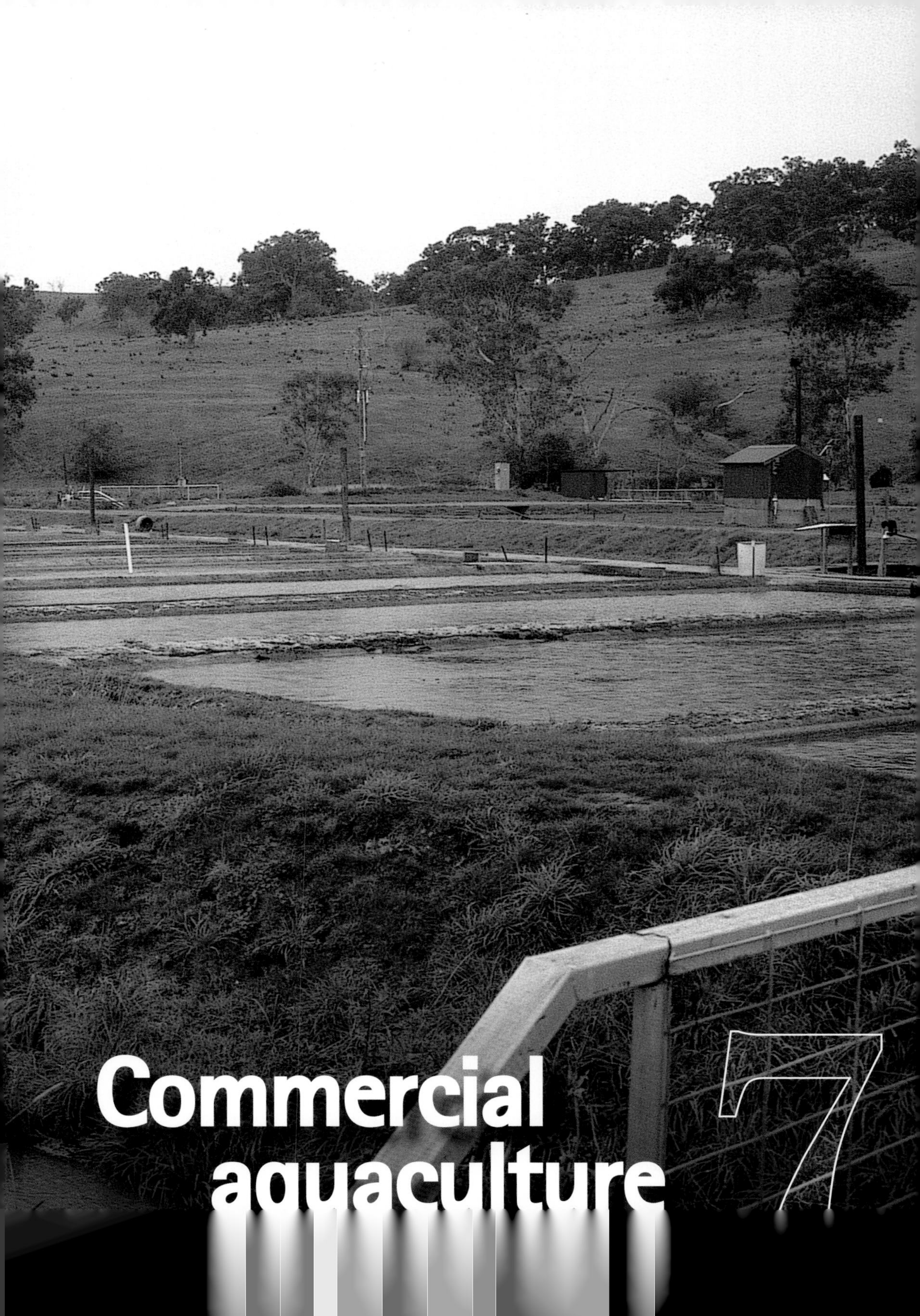

# Commercial aquaculture

# 7

If you are fascinated by aquatic animals and plants, and have a certain degree of skill in keeping and breeding or propagating them, it is an easy step to imagine yourself making a living and a lifestyle based around aquaculture. However, this isn't as easy as it sounds, and commercial aquaculture can even kill your interest in the underwater world!

Raising large numbers of a single species (or even several) for sale once or twice a year is much more tedious and repetitious than breeding small numbers of diverse fishes, and it can also be a high risk proposition which requires attention every single day, sometimes for very long hours. Many growers have been drawn into the field by over-the-top estimates of potential productivity and profitability, though few of these have turned out to be feasible, and over the years most estimates of future fortunes have turned out to be bubbles.

Most people actually involved in commercial aquaculture are well aware of what is achievable at a reasonable price, but there will always be some dubious figures being puffed up by people with a product or a process to sell. Newcomers to the field can easily be talked into investing a lot of money in unproven products or processes, especially if these are described in loving detail in a reputable trade journal such as *Austasia Aquaculture*.

The people who write these articles are aware of the problem. A former *Austasia* columnist commented in the June/July 2002 issue: 'I hear *Austasia Aquaculture* is referred to in some quarters as *Fantasia Aquaculture*. Don't shoot the messenger, boys. We only report what people tell us.' So before jumping in, waders and all, keep in mind that not all that is printed is necessarily true, and that detailed independent research is essential before you consider investing considerable amounts of money and effort on species and systems which aren't going to lead anywhere.

## *An overview of the industry*

Before looking at issues relevant only to freshwater growers, it is useful to give an overall perspective on the considerable growth in aquaculture in the decade since my original book, *Farming in ponds and dams*, was written. In that time, there has been an estimated growth rate (as measured by gross value of production) of around 15 per cent per annum, with the total industry in Australia alone probably worth around $800 million today.

Most of this production increase has been in marine species, of which the fastest growing industry by far was southern bluefin tuna, accounting for around a third of the total growth. Most of the other industries grew at a much slower rate, or in some cases even went backwards. All of this growth is likely to level off over the next few years for a number of reasons, many of which are related to issues of sustainability, and also because the value of some species has fallen or has failed to reach predicted levels.

Using the southern bluefin tuna industry as an example, there is now a quota on how many fish can be captured from the wild and, as little effort has been made to breed this species, there is little prospect of increasing output further. The environmental impacts of the seacages these fish are raised in are pretty appalling, and it is likely that future regulations will require increasingly efficient and possibly expensive ways of disposing of solid wastes, not just allowing them to drop to the seabed below. The profitability of this industry is also very much tied to the presently lucrative but dodgy Japanese economy, and with any downturn in business or increase in operating costs, even this part of the industry may yet end up going backwards.

Other major generators of aquaculture income include the long-established pearl oyster industry, edible oysters, prawns and sea-farmed salmon – all marine. With the generally poor environmental record of the newer prawn and salmonid farming industries, it is becoming increasingly difficult for marine-based industries to expand further (see preface). Their continued viability very much depends on whether markets are prepared to pay premium prices and, as some of these operations appear to be on rocky ground financially, a recession could be enough to put production of such 'luxury' items out of business.

By comparison, the total value of freshwater farming is still relatively small, partly because maintaining high water quality has been a more serious issue here as mistakes can't just be flushed out to sea, and crowding is a much higher-risk proposition. Recent droughts have also depressed freshwater production, as have falling prices for some species as prices adjust to demand on local and world markets.

Even so, most freshwater fishes and crayfish being cultured on a commercial scale are less expensive because the infrastructure required to produce them is also less expensive, and less dependent on high fuel inputs. For this reason, freshwater products are also less dependent on a buoyant economy to keep them selling, and as many of them are also very good to eat they will be more saleable during economic slow times than species that are more expensive to produce.

Nevertheless, there is considerable room for expansion of existing freshwater operations, particularly crayfish, and of a few particularly good quality fishes which are little known as yet. The barramundi industry is a good exemplar of the potential, and in the past few years has become the single biggest income earner in freshwaters (though it is now also being farmed in seacages), more than doubling in value in one recent year.

Much of the future development is likely to take place in the tropics and subtropics, where there is a greater range of previously untested fishes with considerable potential for aquaculture. There is also considerable scope for larger-scale production of established species in southern areas as their needs and feeds become better understood and available. Native freshwater fishes other than barramundi have surged in value by over

30 per cent recently, and this growth is likely to continue more-or-less steadily as the cost of more popular species such as Murray cod stabilises at a lower level, and other high value species such as sleepy cod become more readily available.

The following sections look at diverse aspects of commercial aquaculture, though many other issues of interest to anyone raising anything in or around water have already been discussed in earlier chapters. These include aeration and filtration, protection from predators, use and storage of prepared feeds, crowding and health issues, harvesting, transportation, purging, drought strategies, and of course the basics of pond and dam construction. Other issues will be looked at in more detail under the species and group accounts in later chapters.

## *Types of commercial aquaculture*

Most aquaculturists working on a commercial scale specialise in one or two species, or a particular supply phase for other growers, though there are also a few with much more diverse set-ups. Hatcheries usually raise only fingerlings to be supplied either to other growers, or for stocking farm dams, and most of the larger hatcheries breed and offer a range of fishes suited to their climate.

Grow-out operations raise fish to an edible size, the end goal of most aquaculture production, and supply these in various forms to a range of markets. Although the aquarium industry is a type of grow-out operation, raising small and mostly inedible fish to a saleable size, this is both so specialised and diverse that it is discussed separately (see chapter 8).

The size of the product is a central issue in all grow-out operations, as production costs and grow-out time increase with increasing fish or crayfish size, affecting the number of crops that can be produced. However, it is the market that decides the optimal size for a fish and how large it needs to be grown, and this needs to be investigated long before potentially losing the value of a crop by trying to sell the fish at too large or too small a size.

Asian markets (including within Australia) are likely to accept smaller fish more readily, while the more lucrative restaurant trade requires larger, plate-sized fish, and will sometimes pay a premium for particularly large specimens of prestigious species such as trout. In general, the most readily sold size range for fish is around the 400–600 gram mark, though this varies according to species.

Preparation is the next issue, and the best quality fish are produced by killing, de-gilling, gutting and then chilling as quickly as possible. Scales are usually left on for the sake of

▼ *Holding area for fingerlings soon to be packed at Murray Darling Fisheries.*

appearance, as these can be removed rapidly by the fishmonger once a customer has bought the fish. Larger fish may be scaled (or skinned) and filleted, especially species with high quality flesh such as *Tandanus* catfish, which put many people off with their unexpected appearance.

Crayfish aren't difficult to keep alive for days after harvest, and are usually cooked as close as possible to the time of consumption, as even purged animals start to deteriorate in quality within a few hours of cooking. They would make ideal animals for display in restaurants and fishmongers with live display sections, except that when crowded they may fight and cause damage to each other.

Live fishes fetch a premium price for good quality specimens, but getting them to the relatively few markets available is a specialised business involving a modified water-carrying truck. Any individuals damaged during the harvest are usually killed and possibly filleted immediately, not only because they are less saleable, but also because they are more likely to succumb to infections.

Live-fish suppliers are often located close to cities with large populations where most of their markets are likely to be, for quick access, although with care live fish can also be shipped long distances. The downside of proximity to cities includes much higher land prices to offset reduced transportation costs, and also an increased risk of poaching.

Although most suppliers of grown-out fishes or crayfish operate from their own fish farms, some may also collect or harvest stock from other farms and operations with fewer ponds, a form of cost-sharing which may allow smaller operators to avoid having to buy expensive equipment such as portable cool rooms or trucks with live wells.

In some locations, a number of operations producing similar species for similar markets may work together as a co-operative, sharing the costs of buying and maintaining equipment. This has the further advantage that markets can be supplied with a more even grade of product over a longer time period, ideally with enough farms co-operating that stock can be harvested throughout the year.

Another form of co-operative marketing which has worked well in Western Australia for many years now is extensive stocking of farm dams on many properties with yabbies. These can be harvested by the owners on a casual basis using large traps which will hold the contents relatively unharmed for days, or by a specialist who works from one dam to another through the year. This idea is now being extended into other states, where a multi-water licence allows an operator to stock and harvest dams and ponds on many properties, and could perhaps be extended for other aquaculture species as well.

*▼ Prawn crates used for purging yabbies, with water constantly dripping through from above.*

## Other markets

The best profit on anything you have raised comes from selling direct, whether to the public or to a local restaurant looking for unusual, high quality local produce. This works particularly well in rural areas away from larger city markets, so even a small-scale producer may be able to supply the numbers of fish or crayfish one or two restaurants are likely to need, while the purchaser is able to offer genuinely fresh fish and crayfish which haven't been hauled long distances at considerable cost.

Another extension of this is the fish-out operation, where a sign at a farm gate draws passers-by to catch table-sized fish or crayfish. This can be run in many ways, and may include hire of fishing tackle, and cleaning and dressing services. Usually the fish caught are charged for by weight, as a set charge whether anything is caught or not doesn't encourage return business, and if there is enough traffic through the area a kiosk offering coffee and tea or even light meals can add to the value of the operation. Many other successful smaller aquaculture businesses are also dependent on the tourist industry in other ways, from running their own restaurants to direct sales of value-added products from their ponds.

There is also a diversity of more specialised markets which can be supplied by aquacultured stock, from live fishing bait in the form of small yabbies to breeding water fleas and other live foods for aquarium sales. Some require unusual degrees of skill, for example raising mudeyes (dragonfly larvae) as bait, or salmon primarily for their roe, and it is likely that there are many more such value-adding specialisations than anyone has thought of as yet.

The most sought-after markets are overseas, but exporting is the most difficult goal of all to achieve. Any wholesaler at the other end who is looking for the product you are raising is likely to work on a very large scale, and to expect regular (at least weekly) shipments of a uniform product, hundreds or even thousands of kilograms at a time. This is virtually impossible for any single supplier to provide, and would stretch even the resources of a well-organised co-operative.

The nearest and most accessible markets to Australia are in Asia, which poses an entirely new set of problems. There is the difficulty of low wages and costs which suppliers here must compete against. Secondly, there is a far greater range of aquaculture species available in Asia than here, many of them (from our perspective) vermin which have destroyed aquatic ecosystems worldwide as they escape into rivers and wetlands. Even many of those such as *Clarias* catfish and snakeheads which are indigenous to Asia have already proven themselves to be seriously invasive in other countries.

Many of these species are also highly regarded as food in Asia, fetching premium prices. By contrast, the more limited range of Australian species is not generally highly appreciated overseas, although they have been exported alive and tested in a few countries on that continent. Silver perch are the best example of this, with a peak 1994 production in Taiwan of 500 tonnes declining to 100 tonnes the year after.

Combined with the middle-of-the-range price of around A$5.00 per kilogram at the time, this suggests that silver perch is neither highly regarded for its edible qualities, nor is it likely to be possible to export it from Australia at a price any Asian market would be prepared to pay. If anything, most of the illusory Asian markets are more likely to be competitors – for example, imports of Vietnamese catfish in various forms into the US market have all but brought the long-established catfish producers of the southern states to their knees.

## *Intensive or extensive?*

A commercial pond or a dam must be stocked heavily enough to produce a sufficiently profitable crop; the water quality issues and related problems associated with crowding have been discussed in chapter 3, along with the other basics of intensive systems. Here, it is the cost-effectiveness of intensive versus extensive systems which is considered from the perspective of a commercial grower. Intensive systems are more usually referred to in Australia as recirculating systems, but the two terms are used interchangeably throughout this book (see chapter 3).

While extensive farming can be as simple as just throwing a few fish in a farm dam, intensive systems require considerable knowledge to operate, and are expensive to buy, construct and run. However, they can also be built on much smaller areas of land than other types of aquaculture operation, providing that a reliable source of water is available. This means that viable set-ups can be sited on small properties even within the perimeter of major cities, where land is most expensive.

Although there are a few successful ventures along such lines producing commercial-scale crops within Australia, most of these have been working on their own intensive systems for many years, and sell customised variations of their set-ups as well as market-size fish. As a result, there is little in the way of accurate figures available on how much income derives from the sale of fish, separate from sale figures of systems to other would-be growers. Intensive systems for crayfish are rarely heard of, because provision needs to be made to keep the larger animals from killing or damaging the smaller ones, and the only practical way to do this is in separate containers or cages for individual animals.

Reading between the lines of the many articles which have been published by or about devotees and proponents of various intensive systems over the past decade, some general points stand out. Even the most rabid proponents emphasise that an intensive system can only be viable if it is used to produce high value crops. However, their calculations are often based on the present prices for some species which are popular but still in short supply and will become more readily available in the near future. In other cases, prices may be based on those prevailing in some peak time which may only last for several weeks of the year, and failure to get the crop to marketable size in that period would be a financial disaster.

▲ *A sale tank in a Vietnamese market, holding mixed species including silver perch, barramundi, Murray cod and golden perch.*

Anyone selling a customised intensive system will invariably emphasise that their system works so effectively, and within known parameters, that anyone can operate it once they have had the right training. At the same time, they are also careful to emphasise that projected yield figures are under ideal conditions, which leaves the blame for any underproduction squarely in the court of the operator. This is not dishonesty, but reflects their awareness that the system is likely to function best in the hands of the people who created it, and who therefore understand how to get the most out of it.

The cost of an intensive system purchased from its designer is invariably high – a ballpark figure of around $2 million for a system capable of turning off 50 tonnes or more of saleable fish per year probably being fairly representative. This scale of operation is not excessive, and a consensus seems to be emerging in recent years that this is around the minimal viable size for profitable operation.

For this reason many people interested in intensive culture, or without the land for a larger farm, design their own systems from scratch. Experimental prototypes are usually small, producing only a few tonnes per annum at best, and unable to take advantage of the cost savings which larger operations depend on. This often leaves their designers in a bind if they are undercapitalised – do they abandon a successful prototype, or do they expand it on a much larger and more expensive scale? Unless financial provision is already made for expansion beforehand, most prototype intensive systems will become a waste of money unless they present some dramatic new improvement on existing systems – which is unlikely as they all are being designed on similar general lines!

Although many intensive systems theoretically recirculate water through efficient filtration, all are dependent on at least partial flushing with clean water (5 per cent daily is probably the lowest acceptable figure for grow-out operations), whether from a dam or mains water. If the water is sourced from a dam on the property, could it be used more effectively and inexpensively to farm outdoors? If taken from the mains, flushing water will be a major expense, perhaps even greater than electricity and other running costs – and in the long term, with increasing restrictions on water use, may not be available unless the discharge water is reused efficiently elsewhere.

Finally, no matter how well-designed an intensive system must be, it will be running at a loss if it isn't working at something approaching full efficiency. Costs of setting up, feeding and running the system are high in proportion to the profit levels per crop which can reasonably be expected, so any errors and replacement costs will eat rapidly into profitability. The more mechanised each aspect is, particularly with computerised management equipment, the longer it will take for a novice to understand and use it properly. And the greater the degree of mechanisation and crowding combined, the faster the system will crash if any important component fails, with potential to destroy an expensive crop in a matter of hours in a worst-case scenario.

For this reason, most intensive systems are stocked at something less than their absolute maximum capacity, and include a wide range of backups, from additional generators in the event of a power failure, to supplementary oxygen (rather than just pressurised air) for maximal growth rates. And of course, when a crop is reaching maturity and new fingerlings are being acclimatised at the same time, separate quarantine facilities and good hygiene between ponds are essential to avoid all chance of introducing disease or parasites to the fish about to be turned off to market.

The combination of high costs and high risk in intensive systems is the reason why most aquaculture is still being done in outdoor ponds and dams. These aim to increase production to a semi-intensive level more economically by supplementary feeding, aeration, and perhaps some form of relatively inexpensive, though not particularly effective, filtration. For lower-value species such as silver perch and eel-tailed catfish, semi-intensive systems will remain the only option for large-scale production.

A recent trend in intensive culture which has been gathering momentum is the increasing use of cages and raceways in larger, inland water reservoirs, which use much less technology yet have comparable yields to recirculating systems. Here, the fish are penned in a small, protected area where they can be closely observed, much as in the sea cages which have caused environmental problems through solid wastes building up on the seafloor below.

However, the beauty of these inland versions is that the reservoirs they are in, whether freshwater or saline, can't just be used as a dumping ground for wastes – otherwise the water quality through the whole reservoir would deteriorate rapidly. Instead, cost-effective systems for disposing of solid wastes have been worked on from the beginning, and with time these will hopefully be accepted and applied in marine aquaculture as well.

With trial densities of 80 kilograms of fish per cubic metre working well, these very intensive systems compare favourably with the so-called recirculating systems, but are far more economical to set up and run. As such systems are fine-tuned and expanded onto a commercial scale, they will become a way of value-adding to existing inland reservoirs, offering additional crops from the same water and without the wastage involved in flushing the tanks of recirculating systems.

## *Finance and investment*

Freshwater aquaculture is in a state of active growth in Australia, and although much research on established species has been conducted by various government bodies and academics, the newest and most innovative work is often carried out by individuals or small groups. Innovative doesn't necessarily mean good, however, and the high failure rate of 'start up' operations has meant that it is increasingly difficult to obtain finance from banks or other lending institutions. In turn, undercapitalisation has finished off more than a few businesses which could have become commercially viable if they had the funds to expand enough for economies of scale to kick in.

High risk investments are proverbially linked with high return *if* they succeed, but in aquaculture you can spend years throwing money at a species (or sometimes a system) which will never be viable as a commercial proposition. These are invariably chosen because of their apparently high, per-kilogram value, but in many cases they prove to be poorly adapted to the conditions that a successful aquaculture animal will tolerate. For example, many spiny crayfish are visually appealing in marketing terms because of their large size and ornamental appearance, but their slow growth rates and territorial nature make it impossible to produce them at a price the market would find acceptable.

Market issues are often neglected, especially for apparently popular and expensive species – is there a preferred size range which will minimise costs of production while maximising returns? Where are the markets located, and how much product are they likely to accept before prices slump? All of these issues must be addressed as part of the development plan for a new business, because marketing an aquaculture animal is where *all* income comes from. It doesn't matter how good you are at breeding or raising a species if you can't find buyers for it who are willing to pay a satisfactory price.

It is easy, when carried along by someone else's enthusiasm, to assume that they have already done the basic biological and market research. However, an enthusiastic proponent is just as likely to rush in without noticing an important clue which puts an entirely new perspective on whether the venture is viable. Enthusiasts are also often acting under the pressure of wanting to be first in at the ground floor, knowing that other people are also likely to be looking at the potential of any species with a high market value. Preliminary research should be compulsory – read literally everything published about any species that interests you, and don't waste time reinventing the wheel.

Over the same time you could also carry out small-scale trials in a pond or dam, which may be enough to tell you whether your species of choice can be produced economically in your climate and conditions. This involves not only the price achieved per weight sold, but also how large a volume (or surface area) of water is needed to produce it. For viable species, economies of scale will reduce costs of production later, but you can be pretty sure that an animal that costs twice as much to produce during trials as it sells for is not likely to be worth the trouble!

For the investor, it is essential to know the background of the manager and staff of any planned operation. There are numerous aquaculture courses being offered around Australia, but few of their graduates have all that much in the way of practical skills initially. It could be lethal for new businesses to rely on managers with only academic qualifications, employing them to learn practical skills the hard way through trial and error.

If a venture seems feasible after the investigatory stage, it will still be a matter of years before production on a commercial scale can be achieved. During this time a suitable site with an adequate water supply must be found – there is little point setting up a farm at the opposite end of the continent to where the highest prices are being realised. Dams and breeding ponds will need to be constructed, and of course finance must be available to carry the business through until first production.

Where shareholders are involved, financing will often be staggered over several years, and the timing of calls for additional funds should be specified in advance in the prospectus. The first dividends are likely to be years down the track from the initial investment. Even after the first year or two of successful marketing of a saleable product, all income generated initially may need to be put straight back into the business to make improvements on the original set-up, or to expand if this is necessary to take advantage of economies of scale.

Because of the speculative nature of much experimentation being done in aquaculture, investors need to have more than just sharemarket skills. Informed decisions on the aquaculture potential of untested organisms require some understanding of biology and biotechnology, with more than a smattering of financial sense as well. Although aquaculture investment will become increasingly predictable and profitable as our understanding of many species consolidates, for the moment *caveat emptor* still rules – let the buyer beware!

## *Reading further*

*Austasia Aquaculture* is a bi-monthly magazine reporting on most aspects of aquaculture within the Australasian region, and it also offers an annual trade directory with detailed status reports on the industry and its components. Subscriptions are through PO Box 658, Rosny, Tasmania, 7018, telephone (03) 6245 0064.

For more detailed discussion of a diverse range of issues which should be considered before going into commercial aquaculture, see *Marketing in fisheries and aquaculture* (I Chaston, published 1987 by Fishing News Books) and *Marketing: a practical guide for fish farmers* (SA Shaw, published 1990 by Fishing News Books).

- Aquarium systems
- Going commercial
- Major ornamental fish groups
- Reading further

*Interior of Reef and River, a well-organised aquarium shop.*

# 8 The aquarium: a training ground for commercial aquaculture

Learning to keep and breed aquarium fishes is an excellent and relatively inexpensive way to learn the basics of aquaculture, from water quality to filtration, and also the management of relatively crowded waters. Being able to actually see what is going on under-water makes it easy to spot changes in fish behaviour which can signal breeding or health problems, and once the aquarist is familiar with normal behaviours these skills translate readily to larger ponds or even recirculating systems.

On a larger scale, the aquarium trade has been a significant component of freshwater aquaculture in Australia for many decades, although its proportional contribution to total production has shrunk in recent years as edible species have grown exponentially in importance. Around 14 million aquarium fishes are sold annually around the country, with locally bred stocks making up perhaps half of these, and $6 million of the overall value. The bulk of local production is goldfish, so there is still considerable potential for local growers to produce a much greater proportion of the more exotic tropical species – something that can be done even in the backyard for people looking to supplement their income through a uniquely challenging and diverse hobby.

## *Aquarium systems*

The basic aquarium of recent decades is made of glass bonded with silicone sealant, a strong combination which lasts many decades and is corrosion proof even with seawater. As glass is fairly flexible, this is reinforced with struts along the open top in longer aquaria, and the open spaces may be fitted with cover glasses to keep jumping fishes in, and reduce heat loss in heated aquaria.

An airpump can be used to aerate the water, and is also often used to power various types of filter through an airlift tube, where an upward current is set up by the rising bubbles. There is a wide range of filters available, from simple box filters with an airlift, to elaborate and powerful internal and external filters. Although these are mainly mechanical types, collecting detritus in various types of synthetic fibre or mesh, if not rinsed out too thoroughly when cleaned they will also develop a community of nitrifying bacteria, keeping toxic wastes down.

Undergravel filters are largely a biological filter, with slotted plates drawing detritus into the fine gravel, and water being returned to the tank through tubes at the back of the tank powered by airlifts or by small pump heads. The bacterial community which develops in the gravel breaks down and utilises the more toxic components, while much of the inert detritus which accumulates can be raked out with the fingers, and siphoned away during the partial water changes necessary for best health of the system.

Many plant growers avoid undergravel filters as not all aquatic plants appreciate water movement through their root zone, and use trickle filters (see chapter 3) for biological filtration, or other (often very elaborate) systems for simulating the conditions of tropical rivers. Plants will also take up ammonium directly, reducing the need for other forms of biological filtration, but require adequate lighting for growth. The basic aquarium light is a reflector designed to sit on top of the tank, with a range of fluorescent tube types available from plant-growth types to full spectrum, giving light similar in colour range to that of the sun. Various combinations of these may be used, but all growers agree that intensity is important, and some systems even use halogen lights suspended above the tank.

Coldwater fishes need no heating, but most ornamental fishes are tropical or subtropical in origin, and need some heating in southern areas of Australia at least. This is most easily arranged with an electric immersion heater, generally a glass tube with heating filaments inside controlled by a thermostat which can be set to a range of temperatures. Bubblewrap attached with the smooth surface outwards on the non-viewing sides and over the top of the aquarium will help reduce heat loss; double-sided bubble wrap is particularly efficient.

## *Going commercial*

Sooner or later, anyone with a few fish tanks who is breeding even just one or two species will have to decide whether to keep their hobby small scale, or to expand their operations in a more ordered fashion. If more tanks are to be added, rather than scatter these through the house, which is inefficient in terms of power use and water changing, and can also create mould problems in the domestic environment, the decision is usually made to collect them together in a fish house of some kind.

Often this is a modified garden shed or garage, insulated and with clear panels in the roof to allow some natural sunlight, and a centralised heating system, though individual electric heaters may be used to keep temperatures higher in particular tanks. At this stage, it will also become desirable to breed and sell fish to at least cover operating costs, though usually anyone who has taken their hobby this far will already be raising a wide range of their favoured species.

While you can make a small living from a garage full of fish tanks, stepping this up to a full-time job is a very different proposition and should be approached with caution. The building itself will obviously have to be much larger – the general consensus seems to be that at least 200 $m^2$ of floor space and 50 000 litres of aquarium capacity is needed for a viable tropical fish breeding business.

Good insulation is particularly important, and all aspects of the design of the building should consider this, rather than just retrofitting an existing building as best you can. The walls are often lined with freezer panels – ideally 100 millimetres thick for the walls and, to trap rising heat, 150 millimetres on the ceiling. Thinner panels can be used in an existing building, but the thicker ones are large and solid enough that they can be used to make free-standing walls without need of a separate frame.

The most economical heating is by gas, with aquaria at higher levels being warmer so fishes which are happy at somewhat lower temperatures are kept in the lower tanks. The floor will usually be a concrete slab sloping gently to pipe drains so it remains relatively dry, with a layer of dry, compacted sand at least 100 millimetres thick below it to act as an insulator. Running the air temperature several degrees higher than is required in the tanks is more expensive, but eliminates condensation and mould which can cause human health problems.

Building on this scale can only be done with permits for construction, large-scale water use and disposal, and consideration of access to markets. A hobbyist may be able to sell all the fishes produced to just a few local pet shops, but a commercial grower will need to approach a larger wholesaler and distributor to get some idea of which species are most in demand and at what age or size.

▼ *An unfiltered swimming pool used as a fish pond, with cold-tolerant rainbowfish and various exotic tropicals.*

Keep in mind that many ornamental species are now being raised in ponds in tropical and subtropical areas, so they can be sold at lower prices and may also colour up better than those kept in aquaria all their lives. However, this can also work for southern growers as well, as many so-called tropical fishes will thrive outdoors in the warmer months while water temperatures remain above 15°C. In all such cases, it is essential that the ponds are designed so they don't just overflow during periods of heavy rain, but pass through an escape-proof drainage system so feral populations can't establish in local waterways. For smaller fishes, fine mesh will also be needed to keep fry and fingerlings in as well.

Filtration choices vary widely, but are invariably assisted by considerable water volume changes – up to 10 per cent daily is not unusual. Individual filters (which can be as simple as a box filter with synthetic fibre medium) are often used in each tank, but some operators use recirculation

▲ *A well-organised private fish house large enough to gross a monthly income of around $1000 – note the low-light conditions to reduce summer heat, as few plants are grown here.*

systems with all tanks connecting to main filter banks. This has the potential to spread disease and parasites between tanks, even if all water is passed through an ultraviolet steriliser before being returned to the tanks, and isolated quarantine tanks are absolutely essential for new stock in this case.

Feeding of most types of fry is often done with newly hatched brine shrimp, but smaller fry may need smaller live foods such as rotifers or vinegar eels at first. For economic reasons, the growing fry are turned off onto prepared foods (which are available in various finely graded sizes) as soon as possible, increasing particle size as they continue to grow. Many operators make their own food for larger fishes, often with ground beefheart (the relatively lean heart meat of cattle) as a major bulking-up/cost-reducing ingredient, despite known dangers to fish health if this is a significant part of their diet for much of their life.

Many hobbyists with small fish houses raise both fishes and plants together, using a combination of natural light from clear strips in the roof, and fluorescent tubes suitable for plant growth, though this is not practical in a larger hatchery. Instead, the ceiling may be completely closed and the lighting regime designed entirely around the needs of the fishes – sometimes even using plastic plants for those species which prefer to breed on leaves! Commercial producers of aquarium plants grow their plants in greenhouses rather than in fish houses, with many standing in water but growing emergent, and most larger plant suppliers are located in the tropics and subtropics where heating costs are negligible.

Perhaps the greatest drawback of scaling an aquarium hobby into a living is its impact on lifestyle – many operators complain of the difficulty in getting away for more than a few hours at a time, let alone for a whole weekend. Family-operated businesses of this kind are more flexible, providing that at least one member is happy to stay on-site while the others take time off. Employees are a more difficult problem, partly because highly skilled ones are hard to come by, and also because the business needs to step up in scale to pay an additional salary.

There are two serious problems which need to be addressed to make the production of aquarium (and pond) fishes sustainable in Australia, the major one being the amount of waste water generated. At present, this is often used to irrigate lawns on the producer's property. With an estimated annual use rate approaching 2 million litres per annum for a 50 000-litre operation working on a 10 per cent daily exchange, this will be increasingly perceived as environmentally unsound as water restrictions for other users increase. At the very least, such operations should be required to use their waste water for growing some other useful crop, or work out a sharecropping arrangement with a market gardener or other producer.

A less-easily solved issue is the value of the Australian dollar, which has fluctuated considerably in recent years. At the time of writing it was quite high, so a number of growers were finding it increasingly hard to compete effectively against imports, particularly from Asia. Despite present quarantine regulations, an increasing volume of imported fishes increases the chances of importing new diseases (many of which can't

*Some popular aquarium fishes, from top to bottom at left: rainbowfish* (Melanotaenia), *threadfin rainbow* (Iriatherina), *corydoras catfish,* (Corydoras), *and plecostomus* (Hypostomus)*: at right: swordtail* (Xiphophorus), *discus* (Symphysodon), *dwarf gourami* (Colisa).

easily be detected in just the two-week quarantine period presently required). Stricter quarantine requirements would not only reduce the chances of accidental introductions, but would also encourage more Australian breeders to raise a greater variety of fishes within this country.

## *Major ornamental fish groups*

An important decision to be made is whether to concentrate on mass production of easily bred, popular, but low-value, fishes, or to concentrate on a range of more difficult yet more expensive species to be produced on a smaller scale. It is possible to combine both, but not easy. Some of the major aquarium fish groups are discussed briefly below in terms of saleability and ease of production.

### Native fishes

Rainbowfishes from Australia and New Guinea are the most popular native species worldwide at present, because many are very colourful and they are also relatively peaceful in the community aquarium. These include *Melanotaenia* and *Glossolepis* species, as well as related smaller species such as blue-eyes (*Pseudomugil*) and threadfin rainbow (*Iriatherina*). Although many rainbow species are easily bred and prices of mature fishes are relatively high, they don't colour up until they are quite large. It can take anything from six months to a year to produce a saleable fish, and there is little potential export market except for newly discovered colour variants and species, as they are already being bred on a considerable scale overseas.

Other reasonably popular native freshwater species include gudgeons (though the larger ones can be aggressive), some of the smaller catfishes (though most of these are caught from the wild), and a few of the more striking grunters and other larger species which are kept by a small number of hobbyists despite their aggressive nature. Overall, most popular indigenous fishes are tropical and subtropical, and although many are raised on a small scale by growers in southern Australia, the bulk of the most colourful and easily raised species are produced in ponds up north.

### Goldfish and carp

Goldfish are by far the most popular freshwater fish, partly because they don't need warm water, and also the more common types survive a surprising degree of neglect and poor water quality. There are essentially two types of goldfishes: the tougher and more active varieties suitable for ponds, and the fancier and more delicate varieties, many of which will only survive in an aquarium or indoor pond.

The most active pond goldfishes include comets with elongated orange, white or reddish bodies, and the similarly shaped but multi-coloured shubunkins. Longer-tailed forms swim less effectively, but

▲ *Guppies are one of the more popular and easily bred livebearers.*

all are active and fast enough to avoid most predatory birds, and they aren't worried by winter cold in Australian climates. Many of these are bred by a single producer in Victoria, though with recent increases in the value of the Australian dollar, an increasing number are being sourced from Asia.

Most fancy goldfishes, some of which can best be described as gargoyles with scales, are imported, though there are a few growers within Australia producing small numbers primarily as a hobby. Their often hunched or compressed bodies make them poor swimmers, and cold winter conditions can cause various problems with their distorted internal organs, so most fancy goldfishes must be kept in slightly heated aquaria through the winter months.

A limited market for koi carp also exists in New South Wales and Western Australia, though there are often rumours that these could be banned, in keeping with the policies of all other states. Koi are a large fish and require well-filtered ponds if they are to be displayed attractively, and the few commercial breeders probably already supply all the good quality fish there is demand for.

## Other exotic fishes

Of the many families of fishes which are being bred in Australia or imported from overseas, some stand out in popularity for their colours, unusual shapes and behaviours, and even personality. The major groups are discussed briefly here.

▶ *Shubunkins, a hardy and fast-swimming variety of goldfish suitable for outdoor ponds.*

**Livebearers** including guppies, platies, swordtails and mollies are among the easiest species to keep, and many aquarists started their

breeding career with these as they give birth to relatively large, live young which are easily fed and raised. Most of this group prefer slightly alkaline rather than soft and acid waters, but they are not otherwise fussy, and most will also thrive in outdoor ponds over summer (even in Tasmania), where they develop particularly rich colours on a diet of living foods. However, a few species have also gone feral in subtropical Australia, possibly washed out of ponds into wetlands where temperatures never fall too low for them.

**Cichlids** (pronounced sick-lids) are one of the most diverse freshwater fish families, ranging from tiny and peaceful dwarf species to carnivorous giants a metre and more long which don't necessarily get along with each other, let alone any other type of fishes. Many are surprisingly intelligent, and the combination of diverse body shapes and colour patterns, with elaborate breeding and fry-raising habits, attract a wide following.

Most small-scale breeders have at least a few types of cichlids among their bread-and-butter fishes, whether they specialise in African species from the larger, alkaline lakes, South American giants, or have a personalised mix of smaller and more colourful species from both continents. Few cichlids tolerate colder water in ponds, though many larger species are banned in Queensland as some have already naturalised there.

**Labyrinth fishes** include the ever-popular gouramis, which can breathe air and feed from the surface, as well as the more aggressive and cold-tolerant paradise fish. Gouramis vary greatly in size and colour, though all are relatively peaceful, and the males build floating bubble nests and conduct elaborate courtship rituals, as well as guard their young. This family also includes the well-known Siamese fighting fish, which despite its name is fairly peaceful with all but other males of its kind.

**Catfishes** are actually a multitude of very different families, mostly bottom-dwelling though rarely 'scavengers' as reputed. Many are small and peaceful with diverse 'personalities', so there are few aquaria which don't have at least one or two species. Particularly popular are *Corydoras* from South America which can be produced on a large scale by skilled hands, some including peppered catfish and *C. barbatus* being very cold tolerant (to perhaps 10°C). The other particularly popular group is the sucker-mouth types which help control algae in aquaria, from *Plecostomus* to bristle-nose cats.

**Tetras** include giant types such as piranha, but a great number are small and often brilliantly coloured, and though many are actually quite aggressive they are mostly far too small to bother other community aquarium species. The constant demand for the brighter species, combined with the special attention needed to breed them, has made for a lucrative market for the few breeders with the skills and patience to turn them out on a large scale.

The best known **Cyprinids** (pronounced sigh-prin-ids) are carp and goldfish, but there are also many smaller, more colourful tropical species in this family that are popular for their active, often schooling habits, despite a tendency to be aggressive to other species. Barbs are always in demand, and are fairly straightforward to breed on a large scale, as are the so-called 'sharks' and flying fox – really just elongated carps which can become too large and aggressive for most aquaria with time.

Various other unusual animals are offered through the aquarium trade from time to time, though many of them are novelties which aren't much in demand. For example, blue yabbies are too destructive on plants for many tanks and will catch fish at night, while the axolotl (a type of juvenile salamander) is too dull and inactive to please the average hobbyist for long. Perhaps the most appealing and interesting novelty animals for either cold-water or warm-water aquaria are smaller species of river shrimp, which thrive in a community aquarium where there are no large, aggressive fishes.

Some frogs are kept in terraria, but with increasing restrictions on which can be kept or moved from place to place, this is becoming more of a specialist hobby. Freshwater turtles are protected to some degree in many states, though they are still collected from the wild for sale. Their more colourful and ornamental hatchlings were also collected in the past though they are now completely protected, but if a successful breeder could demonstrate that these could be produced on a large scale in captivity, they could become a viable line for the pet industry.

## *Reading further*

There are literally hundreds of books on aquaria and aquarium keeping, from beginner's guides to highly specialised volumes dealing with a single family or even just one species. For detailed discussion of filtration concepts and some state-of-the-art developments in ecological aquaria, *The optimum aquarium* (K Horst and HE Kipper, first English edition 1986 by Aqua Documenta) is a readily available and reasonably lucid account, though the authors promote their own systems which can readily be duplicated with less expensive components in many cases.

For goldfish, *The goldfish guide* (Y Matsui, original English edition 1981 by TFH Publications) is particularly useful, and *Chinese goldfish* (Z Li, published 1990 by Tetra Press) is perhaps the most comprehensive photographic guide to the incredible, though often hideous, range of fancy varieties developed by their original breeders. Readers interested in Australian fishes should join the Australia New Guinea Fishes Association for their quarterly colour journal *Fishes of Sahul* which features a range of articles from hobbyist level to scientific papers – contact PO Box 673, Ringwood, Victoria, 3134.

*The swimming dam at Dragonfly Aquatics serves many purposes from raising edible fishes (as well as smaller native species), yabbies and ornamental plants, while providing a water reserve topped up from a larger dam against fire events and a frost buffer for avocado trees on its west bank.*

- Limitations of available species
- Theoretical and applied polyculture
- Plankton and building food chains
- Polyculture design and planting
- Reading further

# 9 Polyculture

In aquaculture, polyculture is the raising of at least two different crops (animal or plant) together in the same pond or dam. The aim of the exercise is to produce more saleable product or protein from a volume of water by stocking compatible species which use different foods or parts of the shared environment, so that there is minimal competition and maximum use of all resources. The chinampa systems of pre-historic Mexico, in which islands or floating gardens in lakes were used for crops, were possibly among the earliest aquatic polycultures, if it is true that fishes and other aquatic animals were once harvested from lakeside channels between the raised chinampa beds.

However, the best-documented polycultures are those developed in China, using a number of species of carp. These fishes probably arrived spontaneously in heavily stocked duck ponds from surrounding waters, and only those well adapted to such high nutrient, poor-quality waters thrived. Over a period of centuries, the duck farmers also became fish farmers, learning by trial and error which species complemented each other, and in what proportions they should be stocked for best results.

In part, these polycultures were viable because the carps could be marketed alive, an important consideration in warm climates before the invention of refrigeration. In addition to ducks, which were their primary focus, farmers could also harvest more than six tonnes of fish per hectare anually. With more recent technology and knowledge such yields could be increased to over 20 tonnes per hectare annually through more regular harvesting and grading, aeration and supplementary feeding. Despite this, carp polyculture is on the decline now that better quality fishes – both freshwater and marine – are readily available chilled or frozen.

## *Limitations of available species*

Chinese carp polyculture has long been held up as a model of just how productive polyculture can be, and many people have been inspired to try to develop comparable systems in other countries. Most of the resulting combinations have not proven to be particularly useful or productive in the Australian context, partly because of the restricted range of fishes and other animals we have to work with.

Nearly all of the animals available here are native species, and only a limited range of these are worth raising for food, especially when grown in an intensive situation. This continent is old and most of it is very dry, and when it began its drift up from the polar regions millions of years ago there were very few freshwater fishes aboard. The survivors of this impoverished true freshwater fauna today include salamanderfish, a lungfish, and saratogas, none of which are worth eating, though the latter have a limited market as ornamentals.

The vast majority of native freshwater fishes derive from marine families, and have colonised inland waters as the continent continued its northward drift. Although they have now found their way to most parts of the country, extremes of drought and mostly small river and lake systems mean that we still have only a very restricted range of species – less than the number in the African Lake Malawi alone, though this is only a third the size of Tasmania! Our freshwater crayfish fauna is diverse, but in over 30 years of research only three species with the qualities needed for commercial production have been identified.

A limited range of fishes used in aquaculture overseas has also been introduced, most of which are proven or potential vermin, and some of which have already slipped the leash and are damaging native ecosystems. Trout were deliberately stocked for many decades, and have naturalised so thoroughly that there are very few (if any) colder rivers left in anything resembling a pristine state. Redfin perch are still occasionally stocked in dams by the misinformed, where they breed large populations of stunted, bone-ridden fish with no value for sport or food.

Carp (introduced illegally for aquaculture) have been the most destructive release in warmer temperate waters, while tilapias have the potential to scour the ecologies of every tropical river and wetland. Of these various larger introduced fishes, only trout are legal to raise, partly because all trout farms and suitable stocking dams are located where wild populations are already well entrenched.

As a consequence, nearly all of the very limited range of fishes available for freshwater farming in Australia are carnivores, high on the food chain, with very few species feeding on vegetable matter or plankton on any scale. There is certainly nothing available like the diverse range of Chinese carps which feed on vegetation, minute animals and even the wastes of other fishes, while tolerating water quality which would kill many native fishes overnight.

This point is extremely important: the Chinese carp systems cannot be used as a model for any kind of polyculture in Australia. For a more realistic framework we first need to separate out the confused and often unrealistic expectations which have bedevilled the idea of polyculture before looking at how a functional aquatic ecology is usually built up.

## *Theoretical and applied polyculture*

Although commercial growers have experimented with polyculture in recent years, it has more usually been the province of small-scale growers and backyarders hoping to harvest a greater diversity of foods from a pond or two. Most aquatic polyculture ideas in Australia have been strongly influenced by permaculture, a home-grown synthesis of proven and traditional agricultural practices with a theoretical overlay of broader design issues and biological concepts.

There are some problems with the way these ideas have been joined, particularly in the field of aquaculture. I won't repeat the more detailed criticisms from my earlier book *Farming in ponds and dams* here, but it is important to emphasise that most of these plans and ideas have not been tested even some 30 years later, and that they give a false idea of the productivity of pond systems.

▲ *Silver perch, the closest thing to a widely adaptable vegetarian fish among native species.*

For example, author W Mollison in *Permaculture: a designer's manual* states that 'an intensively managed fish pond of 100 square metres comes close to providing a full protein and vegetable resource' for a whole family, from which '*modest* yields of 300–2000 kilograms of protein ... can reasonably be expected' (my italics). Scaling this up, anyone with a practical background in aquaculture will immediately recognise that such a 'modest' figure as 200 tonnes of protein per hectare is vastly greater than even the best managed Chinese carp polyculture could achieve – yet Mollison claims that even higher yields could be achieved just by adding aeration!

Not surprisingly, most permaculture people who have experimented with aquaculture have started out with totally unrealistic expectations, and little notion of water quality issues. Reports of such failed experiments as throwing rotting fruit into a silver perch pond to drown fruit fly were not uncommon a decade ago (the fish died). Other recommended but undesirable and sometimes bizarre practices, such as building compost bins in shallow water, breeding fruit flies on floating rafts, and housing pigs, ducks or chickens over the pond so their manure will fertilise the system, are also to be avoided.

Beyond these fundamental issues, many permaculture ideas are founded on biological principles applied sometimes inappropriately. For example, the same manual suggests that productivity of a pond could be increased by including multi-rayed bays around a central pond. What is not considered is that the living space for the larger, edible species is reduced in such a design for any given surface area, while the convoluted shoreline encourages weedy plant growth and provides more protected habitat for undesirable predators such as mudeyes. It is not for nothing that nearly all aquaculture systems in the world, including traditional ones, lay ponds out in simple circles and rectangles.

▶ *Green water this rich in colour suggests a serious nutrient surplus. The sedge is* Bolboschoenus caldwellii, *a native species with edible, nut-sized tubers.*

On the positive side, this manual remains a valuable and often stimulating source of ideas for placing aquaculture in a broader perspective, whether on a small-scale homestead or a large commercial farm. Among other virtues it promotes an integrated approach to the use of water not only for aquaculture, but also in the context of fire protection, frost control and microclimate modification for plants which are marginal in a colder climate, as well as human habitat.

By contrast, commercial experimentation with polyculture is

generally confined to simple combinations of two or sometimes three species, and given the limited range available the combinations tend to be the predictable and obvious ones. Some pairings were obviously doomed to failure – for example, raising rainbow trout and yabbies together. While neither species will bother the other if stocked at appropriate sizes, and they feed in different ways, their preferred temperature ranges are very different with trout needing deeper, colder waters to survive a hot summer, while yabbies grow best in shallow, relatively warm waters. The two species *could* be combined in specially designed ponds which offer both habitats during the warmer months, but whether the combination would be ideal for either is a moot point.

Other combinations are more practical and look to have more of a future, particularly silver perch and yabbies, or another freshwater crayfish suited to local conditions. While silver perch will eat small yabbies, they are far from being an efficient predator, and the two species will grow up together with few problems. Indeed, the limited predation will allow the remaining yabbies to reach a greater size in a shorter time, free of competition from an excess of siblings.

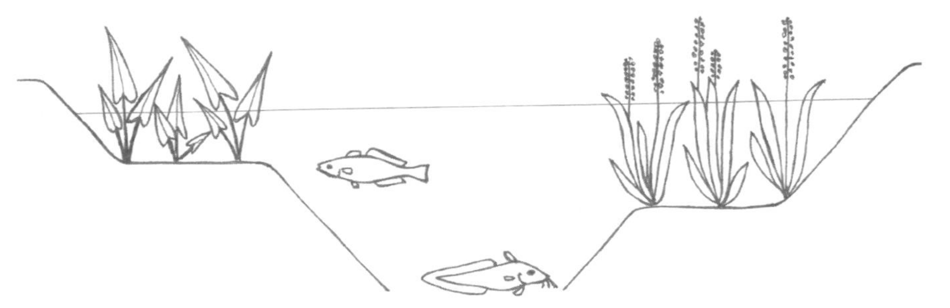

*Cross-section of a permanent polyculture pond, showing planting shelves at different levels with a deeper central area for fish.*

Silver perch are also the basis of most other large-scale attempts at polyculture, because they are the closest thing to a vegetarian and won't eat small fishes – probably because they can't catch them! Combinations of silver and golden perch have also been tried with some success, at a ratio of ten silvers to each golden, though this seems a little high to me, and 20:1 might be more appropriate. Eel-tailed catfish are also compatible with most other larger fishes, though they can develop a taste for smaller fishes, so the adults should ideally be removed before stocking new fingerlings of other species. An active predator on freshwater crayfish, catfish will eat nearly all of their young so the adults are left to grow to a greater size.

It should be noted that most attempts at commercial polyculture are in extensive ponds, with stocking rates moderate at best. This makes it hard to quantify whether there is any actual advantage to polyculture, as any apparently high total yields may just be a reflection of uncrowded conditions and abundant resources. Plants have also been experimented with in such ponds, particularly as forage for freshwater crayfish, and some of these are discussed in chapter 12.

One potential field in which polyculture ideas could be more widely applied in Australia is in flooded rice fields, where an animal crop could often be raised in the same waters. This is done with shrimp and fishes in Asia, though the saleability of the product often depends on a greater acceptance of smaller fish as plate-sized. However, rice fields could also be used as a nursery area inaccessible to wading birds and cormorants, shifting the partly grown animals to other waters for their final grow-out phase when the paddies are drained to allow the rice to ripen.

## *Plankton and building food chains*

A polyculture pond needs a rich and diverse ecology of smaller and miniature organisms to provide feed for larger species, so it needs to be at least reasonably biologically mature before being stocked. Plankton are the basic natural food for fish fry and even some larger fishes, and these are separated into two primary groups. The green, photosynthetic ones which fix solar energy directly are called **phytoplankton**, and include a diversity of single-celled, animal-like species as well as a range of true algae. These (and also some types of bacteria) are the food source for **zooplankton**, from the tiny and complex rotifers which are an important food for fry, to larger animals from water fleas to copepods which are fed upon by a wide range of fishes.

Phytoplankton blooms (green water) can be encouraged in newly established dams by adding moderate amounts of straw and allowing it to rot down naturally – this is often done in ponds which are drained between harvests. Supplementary nutrients in the form of blood-and-bone or small amounts of manures can help in nutrient-poor areas, but must be used cautiously as there is a fine line between a beneficial bloom and a serious water quality problem. Soupy green water can be particularly lethal in a still water pond on hot nights, when the phytoplankton are using oxygen instead of producing it.

If the pond or dam is kept in constant production because harvesting is irregular and carried out over long periods of time, various types of small predators such as backswimmers may build up in numbers, and plankton will become a relatively scarce resource, even if the green water is kept up at an appropriate level. Such ponds aren't suitable as breeding ponds because many plankton predators are also partial to newly hatched fry. However, if the pond is large enough, and is stocked with fish or crayfish that breed naturally in still waters, it is likely that enough young will survive to maintain population numbers.

To overcome some of these problems, commercial (monoculture) ponds are often drain-harvested when the crop is ready, allowing them to dry out for a while before re-filling and re-fertilising, which encourages a new bloom of highly nutritious plankton while discouraging their predators. Another variant on this (particularly in crayfish ponds) is to drain and re-fertilise only every few years once production shows

▲ *A white-faced heron is finding it difficult to fish along this steep, heavily planted dam bank.*

signs of dropping off, using nursery ponds for any large-scale restocking, or bringing in fry from specialist breeders as required.

In dams that are never deliberately emptied, the emphasis is usually on encouraging permanent populations of other (larger) fodder animals, including smaller fishes and shrimps. These can be added once enough plankton is present to support them, and may help to control the build-up of plankton predators, so that something approaching a natural ecological balance develops between the different species with time. Fingerlings of larger species such as silver perch can also be added early, but true predators such as golden perch or barramundi should not be stocked until substantial populations of fodder animals are present.

## Polyculture design and planting

Many aquatic polyculture systems are designed to produce crops of both animals and plants, although there is little chance of raising edible-sized animals in most smaller ponds (see chapter 3). Any such mixture of plants and animals is usually selected for staggered or intermittent harvests rather than all at the same time, as this would require draining of much of the water in the pond. If fishes are to be included, a deeper sump (which may make up most of the water volume if the pond is dam-sized) is necessary, as few of the species available will grow well in shallow waters only.

Plants are catered for by including shallower areas, often in the form of level shelves covered in topsoil, set at different depths depending on the species to be grown here. Shelves can be of any width and area, and are arranged at various depths around the fringes of the main water body – making allowance for evaporation, which increases in direct proportion to increasing water surface area.

As many productive aquatic plants require a drying-out period to form good crops of tubers or seed, any shelf they are to be grown upon needs to be at a suitable depth that will be left uncovered by evaporating waters at the right time. For example, in an area where about 40 centimetres of water disappears over a four-month period, a shelf at this depth would be at about the right level to allow rice plants to dry out and ripen their seed. In southern Australia, the rice harvest would be in autumn, while in the tropics it would fall not long after 'the wet'.

Shallower shelves such as these also provide browsing areas for crayfish and shrimps, while mosquito wrigglers are kept under control by fish if these have space to forage between plants. Shelves at different depths allow for growing a wider range of aquatic plants, and continue

*▼ Domestic ducks in large numbers can foul water, and may also undermine pond banks by dabbling along the waterline.*

to provide suitable habitat for crayfish as the shallower areas are exposed by evaporation.

A simple set-up like this will potentially cater to a wide range of species, both plant and animal, but other issues for enhancing production should also be considered. A combination of aeration and supplementary feeding can increase the carrying capacity of the pond at a relatively low cost, but it is essential to be able to keep track of how much food is being consumed, especially when water temperatures are falling.

Natural feeds may need to be stocked or augmented, depending on the species of larger fishes to be raised. Shrimp are readily eaten by most predatory fishes, but should be introduced at least a few months in advance of predators to give them a chance to build up in numbers. Smaller fodder fishes should also be introduced well ahead of piscivorous predators. Among the most prolific and readily established fodder fishes are the little *Hypseleotris* gudgeons, of which there are a number of species over much of mainland Australia which breed readily in ponds.

Other smaller native fishes such as rainbowfishes (various *Melanotaenia* species) which will breed in still waters can also be used as fodder fish, but there is no point introducing any fishes that need running water or that migrate to estuaries to breed. It is essential to be able to recognise and avoid the introduced plague minnow (*Gambusia holbrooki*) which breeds like flies in most ditches, ponds and creeks in many parts of this country, and will destroy fish fry as well as attack the fins of young fishes.

There are no freshwater snails worth raising as food in Australia, mainly because they are too small. However, freshwater mussels are often recommended, despite the parasitic stage (known as black spot) which their young pass through buried in the flesh of fishes. Only mussel species which tolerate still waters will thrive in ponds, so random harvesting from flowing streams is more likely to create a water quality problem as they die off. Even if mussels are successfully introduced, their main liability as a cultured animal is lack of flavour – the texture is pleasant enough, but they taste of nothing but the herbs and spices they are cooked in.

Some frogs are good eating, but these will never form a productive part of an extensive system in Australia. As they are mostly now protected for good reason, they are best just encouraged as insect control, though their tadpoles may be an important feed for some carnivorous fishes at some times of year. Freshwater turtles are also protected, and as these compete with predatory fishes for fodder fish and also eat small crayfish, should just be netted and returned to your nearest creek if they walk in overland – low fences tilted outwards will help to keep them out.

Ducks should be included in very moderate numbers only, and although their wastes will give green water organisms a boost, in excess they can cause massive water quality problems which completely alter the ecology of a pond, rarely for the better. Breeds such as Campbells and

Runners which have been selected mainly for egg production should be kept locked up in the mornings, which is their laying time, as these high strung types may otherwise lay in the pond, combining lost production with water quality problems.

## *Reading further*

For a detailed discussion of the Chinese carp polyculture systems see *Aquaculture in Taiwan* (L Chen, published 1990 by Fishing News Books). The most readily available account of chinampas is 'The chinampas of Mexico' in *Scientific American* (MD Coe, 1964, 211: 90–98).

For various working models of raising fishes, prawns and other animals among rice see *Rural aquaculture* (editors P Edwards, DC Little and H Demaine, published 2002 by CABI Publishing), *Rice-fish research and development in Asia* (editors CR de la Cruz, C Lightfoot, BA Costa-Pierce, VR Carangal and MAP Bimbao, published 1992 by ICLARM), and *The power of duck: integrated rice and duck farming* (T Furuno, published 2001 by Tagari Publications).

*Barramundi – the greatest success story to a date among native freshwater fishes.*

- Cold-water species
- Southern species preferring warmer waters
- Eels
- Tropical and subtropical fishes
- Reading further

# 10 Fishes

The relatively small range of freshwater fishes suitable for aquaculture in Australia was discussed in the previous chapter, while the more important species being cultured at present or likely to be worthwhile prospects for the future are looked at here. Nearly all of these species are known to be salt tolerant to some degree, which is not surprising given that they are all descended from marine groups, and that some must even return to estuaries or the sea to breed. This could be a useful feature of their biology given the rising salinity levels which are becoming an increasingly serious environmental problem through much of Australia.

Some poorer quality or more difficult species which could be raised in ponds and dams have been omitted because they are unlikely to be stocked when better quality species are readily available. These include spangled perch (*Leiopotherapon unicolour*) which are too small, bony and aggressive to be either profitable or mix well with other fishes; saratogas which are too large, bony and aggressive; and bony bream (*Nematalosa erebi*) which are just too bony and fetch a minimal price on the market. Introduced vermin such as carp (*Cyprinus carpio*), tilapia (particularly *Oreochromis mossambicus*) and redfin perch (*Perca fluviatilis*) were discussed briefly in previous chapters, and don't need to be considered further here.

Before looking at the fishes themselves, it is worth commenting on the nature of aquaculture stocks. Such fishes are selected for high food conversion rates on artificial diets, tolerance of crowded conditions and less-than-optimal water quality, their ability to breed in captivity, and to some degree for disease resistance. Inevitably, long-cultivated aquaculture stocks become different from wild stocks, but breeding from a *small*, randomly chosen population without selecting for those with the best attributes for culture will usually lead to inbreeding, not improvement.

This is less of a problem with freshwater crayfish, as most growers select their own new generations of breeders, and numbers on even the smallest commercial farm are large enough that inbreeding problems are slow to develop. However, most fish fingerlings are supplied by a relatively small number of specialist hatcheries at present which don't necessarily have access to the very best of each new generation for further breeding. Some specialist grow-out operations are coming to see the benefits of selecting their own breeders from the fastest growing and most vigorous fishes they raise, and some are experimenting with crossing vigorous strains of fishes from different sources with promising results.

If growers and hatcheries come to work together more closely, so that particularly fine specimens from the many millions of cultured fishes raised each season were returned to the breeders, similar improvements could be made across all of freshwater aquaculture. However, it should be emphasised that improved aquaculture stocks are (literally) a very different fish from those used for restocking natural waters. The latter should be bred from as wide a wild gene pool as possible, and derive as directly as possible from local, wild, unselected fishes (see chapter 5).

Culling and grading are related issues for grow-out operations, particularly for some of the more aggressive species. Culling is the disposal of smaller and slower-growing fish which are likely to continue to grow so slowly that they won't be ready for market until many months after their larger siblings. As they will then be occupying waters which may be needed for the next batch of fingerlings, or holding up the drying out and fertilising of a pond, it is often more economical to dispose of them while they are still small – rather than wasting more food on the runts of the litter.

At the other extreme, cannibalistic fishes such as Murray cod, barramundi and sleepy cod will cull their own smaller siblings, so that in extreme cases you may end up with just a few large fish from a starting stock of hundreds. As these all have big mouths and stomachs, any single fish which can swallow a smaller sibling gets a head start in growth, and can thin the ranks even faster as it grows further ahead of them. These predators can still be successfully raised intensively because they are easy to fish out of a confined pond, and can be regularly graded into groups of a similar size which can't swallow each other.

## *Cold-water species*

The best known cold-water fishes cultured in Australia have been introduced – various species of trout and salmon. Of these, **brook char** (*Salvelinus fontinalis*) has much more specific environmental needs than brown and rainbow trout, and is unlikely to thrive in a pond or dam. **Brown trout** (*Salmo trutta*) were introduced as an angler's fish rather than as a culture species, and though fingerlings are occasionally available these are mainly intended for stocking larger dams sparsely to produce large, semi-wild fish for sport.

Brown trout are less tolerant of warmer waters and still conditions than the long-farmed **rainbow trout** (*Onchorhynchus mykiss*), which is the major culture species in this country. These are a good looking fish with pink striping and black spangling over their silver bodies, but their reputation as a fine edible fish is somewhat exaggerated. Though particularly good to eat if cooked immediately after capture, or smoked by an expert, freshwater rainbows are by-and-large fairly bland, and are largely dependent on their reputation as the classic sport fish of the English upper classes for their market niche.

Both browns and rainbows will also run to the sea when the mood takes them, and such fish are more strongly flavoured. Their preferred temperature range is somewhere between 10°C and 20°C, and temperatures below this range won't cause them any harm, even if the water surface freezes over. High dissolved oxygen levels are essential, but won't be enough to keep them alive if the water temperature climbs much above their preferred range.

▲ *Rainbow trout are the main non-native fish cultured in Australia.*

Rainbow trout can easily be raised in large dams, which should be at least 3 metres in depth so that the fish have a deeper, cooler reservoir to escape into during hot weather. Stocking rates vary according to the size of fish to be harvested – anything from 100 to 300 fingerlings per hectare will allow individual fish to reach a good size, though some supplementary feeding is likely to be needed at the higher densities. Removing the larger fish regularly will allow smaller siblings to grow faster, and this is easily done by angling as cultivated trout are a bit brainless, and are popular with fish-out tourism operations for that reason.

The commercial markets for rainbow trout are fairly crowded and competitive, so profitability is dependent on efficient use of feeds and water. Specifically prepared pellet feeds are readily available for trout, giving good conversion ratios, and most farmed fish are raised in raceways with constant water interchange provided by a nearby stream. With increasing concerns about the impact of trout farms on the downstream environment, it is unlikely that many more will be approved in the future – but then, there is no reason to suppose that any more are needed to fill domestic needs, when the export markets are already very competitive.

**Atlantic salmon** (*Salmo salar*) and **quinnat salmon** (*Onchorhynchus tshawytscha*) both spawn in fresh waters, and fingerlings are usually raised in fresh water as well. Most of those raised are Atlantic salmon, destined for sea farms, with some quinnat being used to stock recreational fisheries in landlocked lakes. The one competitive edge a freshwater hatchery has over sea farms is in the production of salmon eggs, which are a highly priced delicacy, though it is likely that this niche market is limited and present-day production is unlikely to be expanded greatly in the future.

There are very few native freshwater fishes in southern Australia which are large and tasty enough to be worth culturing, let alone breeding in farm dams. **River blackfish** (*Gadopsis marmoratus*) is a well-flavoured species found through much of the cooler and more southerly parts of South Australia, Victoria and New South Wales. Most don't grow much larger than 30 centimetres though exceptional fish can be twice this length – the largest forms are found south of the Great Dividing Range and in Tasmania.

This species feeds by night on a variety of animal foods, and will breed in hollow logs or pipes. Individual fish don't move far from their home hollows, and providing abundant shelter in a dam will allow greater population densities, though the fish themselves may be smaller with more competition for the live foods available. There are no guidelines on stocking rates available, but given enough shelter anything from 100 to 200 fish per hectare should be feasible.

Although larger, southern forms have long been transplanted outside their native range and establish readily in farm dams with adequate shelter, their commercial potential appears to be limited by slow growth and their aggressively territorial nature. The **two-spined blackfish** (*G. bispinosus*) is usually too small to be worth stocking, and has a much more restricted and higher altitude range.

**Estuary perch** (*Macquaria colonorum*) is a largely coastal fish of south-eastern Australia, while the closely related **Australian bass** (*Macquaria novemaculeata*) is more of a warm-water species, extending southwards into eastern Victoria. Although both are highly regarded as angling fishes because they are wary and put up a fight when hooked,

▼ *Golden perch are a large carnivorous species widely used in ponds, but adults will eat fingerlings of any species introduced later.*

and fingerlings are easily raised for restocking wild fisheries, neither species grows particularly well in dams so it is unlikely that there will ever be much of a market for them.

## *Southern species preferring warmer waters*

The Murray–Darling river system extends across a substantial area of south-eastern Australia, and although there are cold-water species such as blackfish in the higher and cooler reaches of the Murray, most fishes from these regions need warmer conditions to breed. Other populations of the same species are more isolated in various coastal streams from around Sydney northwards. All of these fishes grow faster in the warmer parts of their range, so it is likely that commercial production will be most economically viable in southern Queensland, and northern or western New South Wales.

Even within this area, production takes longer in the more southern areas, with a table-sized silver perch which can be raised in 18 months in Queensland needing two years or longer in cooler climates. This disadvantage can be offset to some extent by stocking with larger fingerlings (not necessarily much more expensive) just at the time when water temperatures climb into the preferred range for each species.

All these fishes are much more cold tolerant than their favoured temperature range would suggest, and can also be stocked as food fishes right down into southern Victoria – just don't expect to make a living from them this far south. As a rule of thumb, expect any of the Murray–Darling species to take between 50 and 70 per cent longer to reach an equivalent size in colder climates, though again this can be offset to some degree by stocking with larger fingerlings at the beginning of the growing season.

**Silver perch** or **bidyan** (*Bidyanus bidyanus*) is the best known and researched of the edible southern natives, and although it is a good quality eating fish if grown uncrowded in clean water, or purged well after being raised in more crowded conditions, interest in it is mainly confined to Asian markets within Australia. Despite this, the prospects for developing an export industry to south-east Asia are small (see chapter 7).

Part of the problem is the unrealistic expectations raised by the intensive studies published on this species in the mid-1990s. These set the groundwork for the present-day industry, attracting many smaller investors who set up ponds without much consideration to how they were going to deal with preparing and marketing their fish. As a result, competition rapidly dropped the comfortably profitable prices being paid initially, and many of the smaller, undercapitalised and poorly planned ventures were also selling unpurged fish.

No one who tasted one of these musty, off-flavoured fish would want to repeat the experience, and it is likely that this factor alone has kept the industry relatively small, with only a few hundred tonnes being sold per annum. Although live silver perch that have been properly purged are available in Asian markets and some restaurants, even here this species has the disadvantage that it is the most easily stressed of the native fishes regularly shipped alive.

As a result of these problems, the silver perch industry is in a bind. Prices are low enough that profit margins are small, and many smaller growers (who have also been plagued by drought in recent years) have given up. The larger operations still operating are better set up and their marketing is more organised, but to develop the industry further they need to be able to cut costs further. The most obvious way to do this is by feeding with pellets which are specifically formulated for this species, rather than for others, as food conversion figures quoted to date aren't particularly impressive. Yet the small size of the present industry means that it is not worth the effort of any of the larger feed suppliers to research more specific and economical feeds.

Commercially, silver perch are crowded at thousands of fish per hectare, but for farm dams and domestic purposes stocking should be kept between 200 and 500 per hectare, assuming that supplementary feeding will be available at the greater densities. At these higher densities it will also be necessary to purge the fish for anything from four days in summer, to four weeks in winter – more usually between one and two weeks. For domestic production, this can be done simply by dropping up to a half-dozen fish into a clean pond of at least 5000 litres, and changing 10 per cent of the water daily if the fish are left for more than a week.

Silver perch grow best at temperatures between 16 and 30°C, and much below this range they don't feed at all. A thermometer isn't needed to tell when they are likely to eat as this is a mid-water to surface schooling species which becomes active once temperatures are to its liking, and hungry fish can be seen picking insects off the water's surface on warm afternoons. This is also an excellent time to catch them on a small hook baited with a river shrimp, and kept up near the surface with a float. They will also eat most other smaller invertebrates and softer plant matter including algae, making them an ideal candidate for polyculture.

**Golden perch**, **yellowbelly** or **callop** (*Macquaria ambigua*) is the second-largest of the big predatory fishes of the Murray–Darling being cultured and, while not quite as cannibalistic as its relative the Murray cod, will eat anything it can fit into its mouth, including its own kind. It is an attractive, deep-bodied species varying from green to gold depending on source of stocks and water clarity, with well-flavoured, moist, slightly oily flesh when taken from clean waters. However, this species seems particularly prone to off-flavour when crowded or raised in muddy waters.

Stocking golden perch at more than 100 fish per hectare is unnecessary, and they are likely to cull each other even at that density – only 20 to 30 per hectare should be stocked if they are to be raised with other fishes such as silver perch. As a large, carnivorous fish that doesn't take prepared foods readily, golden perch is not ideal in polyculture, but their fingerlings can be introduced once other fishes have already had a few months head start in growing time. They will feed mainly on any insects, crayfish and fodder fish in the pond or dam. All mature golden perch must be removed before restocking with fingerlings of any fish species, including their own.

**Murray cod** (*Maccullochella peelii*) is the largest of Australia's freshwater fish, and younger ones make good eating if they are taken from clean waters, or purged before use when raised intensively. It has long been stocked in dams, with unpredictable results, and mature fish will readily breed in hollow logs and similar places underwater.

As a large and aggressive carnivore which will eat its siblings or any other fishes, this species is not well suited to polyculture. In the past it has been suggested that 100 fish per hectare was a suitable stocking ratio, but often only two or three large fish would be caught several years later – even without any help from cormorants in reducing numbers. A lower stocking rate of a dozen or two per hectare is more likely to produce consistent growth so that they don't have the chance to thin each other out.

Despite this, Murray cod are being successfully cultured in recycling systems, where the growing fish are readily fished out and can be graded regularly so the larger ones don't eat their smaller siblings. No special diets have been devised for this species yet, though it is reported to grow well on barramundi pellets. While prices remain reasonably high it is obviously a commercial proposition, even if raised in intensive conditions, but as it becomes more readily available profitability is likely to fall.

As an iconic Australian species, Murray cod has the potential to become a success on the scale of barramundi, especially if specialised feeds and proven raising techniques become widely available. Unfortunately, it is also reasonably well regarded in Asian cuisines and young fish have been exported to China, which is likely to close off export markets if it is well accepted and bred there. Future markets for locally raised fish are

◀ *Murray cod are mainly raised in recirculating systems where they can be regularly graded, so the larger ones don't cannibalise the slower-growing specimens.*

▲ *Eel-tailed catfish are good to eat, but their unusual appearance, and large proportion of head to fillet have impeded their use as a commercial species.*

therefore likely to be limited to Australia, and the cost of the intensive systems and regular grading needed for profitable production could bring the local industry down before it can really take off.

**Trout cod** (*M. macquariensis*) is a smaller relative of Murray cod, but is presently regarded as endangered, as it has disappeared where its range overlapped with that of its larger relative in some of the cooler waters of the Murray. It is being restocked and encouraged within that range, and it is possible that surplus fingerlings will become available for stocking farm dams in the future, in which case its culture is likely to be comparable to golden perch and Murray cod. **Eastern cod** (*M. ikei*) from northern coastal New South Wales is also regarded as endangered, but with successful restocking may also become available for stocking dams within its native range in the future.

**Freshwater** or **eel-tailed catfish** (*Tandanus tandanus*) is a good edible species which is not widely cultured because its unusual appearance puts many people off if it is marketed whole. The head is also very large in proportion to its length, not leaving a lot of meat once even a quite substantial specimen has been filleted. As this species is also an active carnivore which prefers to prey on fine food animals such as crayfish, it is unlikely to become a commercial proposition.

However, a few catfish added to a polyculture will use resources other Murray–Darling fishes don't have much use for, and will do an excellent job of preventing crayfish from over-breeding so that their average size falls away. Stocking rates have been recommended at 100 to

200 fingerlings per hectare, but in a polyculture situation 50 would be a more suitable rate.

Catfish eat mainly live foods including most types of small animals, and can even learn to chase and capture small fishes. They will sometimes enter traps baited with a crayfish, and can be trained to feed around a jetty where surplus crayfish are occasionally broken up as berley, so they are then easy to catch with just a baited hook and a reasonably strong line. They should be skinned before eating, using a dry cloth to hold the killed fish and avoiding their spines which are not only sharp but also add a nasty, stinging irritant to any wound.

This is one of the few inland fish which will breed readily in a dam, making a raised nest of pebbles for the eggs to be laid on, and the male guards the eggs and very young fry. The water needs to be fairly warm to trigger spawning, and must not fluctuate much in depth – even rapid evaporation can be enough to put the would-be breeders off.

## *Eels*

Eels have been separated here from all other fishes, partly because of their very different biology, and partly because the Australian species range from South Australia to the tropics so they cannot be easily pigeonholed within the other general categories used in this chapter. Unlike most native freshwater species, our eels breed in tropical seas, with their offspring (at first leaf-like and clear, later turning into glass-like miniatures of the adults) following currents back to the eastern coastline and changing into little eels (elvers) before they move upstream.

There they remain for decades in some cases, before this next generation becomes large enough to head out to sea. As a result of this unusual lifecycle, the greatest obstacle to eel culture has been obtaining a regular supply of elvers for stocking purposes, with all of these being captured by various types of trap as they move into estuaries. Not much has been done with eel culture in Australia, and the elvers are generally exported or stocked into land-locked lakes, from which they are harvested with a type of trap called a fyke net.

However, a New Zealand venture has recently successfully spawned mature short-finned eels, captured on their way to the spawning grounds, by controlling various factors from light regime to salinity, as well as ovulation-inducing hormones. Methodology for raising eels from glass eels is already well established, so if the newly hatched eels can be raised to glass eel stage the cycle could potentially be closed, though hatcheries would continue to be dependent on wild populations for spawners.

Eels are intensively cultivated in Japan, where high quality diets have been worked out and the business has become a science. Regular grading is necessary to prevent larger animals from eating their much smaller siblings, though males and some other fish often remain small despite an abundant diet, and it may be more appropriate to cull these at an early age if they are markedly slower growing.

Similar methods could be applied in Australia for local species, assuming that it is economically viable – the problem being that eels have been dropping both in value and in production volume worldwide despite what appears to be a steady demand. It may be that the Australian species compare unfavourably with those raised overseas, as they are reported to be somewhat lower in oil content.

Eels don't appeal to every palate, and many people are put off by their snake-like appearance, but those of us who like them regard them as one of the finest edible fishes. Their flesh is rich and oily, with a smooth texture which is enhanced by smoking, though they are also good marinated and grilled. The best quality eels are harvested in their 'silver' stage, when they are changing colour and preparing to return to the sea, though they are good to eat at all stages.

Laws and attitudes to eels vary greatly between states, cultures and even small country towns in Australia. Many aquaculture producers regard eels as vermin rather than product, as these skilful night hunters can make massive inroads into populations of young crayfish as well as smaller fish fingerlings. Despite this, the laws remain confused for eels and in some places it is still illegal to sell an eel that has been marauding in your ponds, but perfectly all right to kill it and leave it on the pond bank to rot. My advice is to develop a taste for eels, and eat them yourself so the legal issue doesn't arise.

In coastal areas where eels are a problem, the elvers move inland during wet weather, wriggling through shallow flowing water to ponds and dams. They are reasonably easy to discourage simply by running an overflow pipe from each pond or dam so that its end is at least 30 centimetres above the ground – as the elvers can't jump, they are unable to recognise where the water is coming from if it falls from above.

Two of the four native eels are relatively small and localised in distribution within the tropics, and are unlikely to find a place in aquaculture. The **marbled** or **long-fin eel** (*Anguilla reinhardtii*) is found virtually along the entire eastern coast on Australia, and reaches around 22 kilograms, though such monstrous specimens are not common. It is not particularly popular as a food fish in Australia, but is being exported to Asia where it is much in demand. The **short-finned eel** (*A. australis*) generally reaches only around 3 kilograms and a bit over one metre in length, but is the preferred species for domestic consumption because of its somewhat firmer flesh.

## *Tropical and subtropical fishes*

Despite the more diverse range of larger, edible fishes in more tropical areas, surprisingly few have been tested or bred for aquaculture on any scale. **Barramundi** (*Lates calcarifer*) is the biggest success story in freshwater aquaculture over the past decade, with an annual production of around 3000 tonnes, though the largest single farm for this species is now a seawater facility in the Tiwi Islands of the Northern Territory.

*▲ Jade perch put on fat readily when cultured. Their marketing name was chosen to replace the original – Barcoo grunter – for obvious reasons.*

However, many farms continue to operate successfully in fresh water including using cage cultures in enormous freshwater reservoirs, and even in southern Australia where warm artesian waters are available.

In part, the success of this fish can be put down to its ability to shift between freshwater and the sea, so it is also found through much of coastal Asia. It is also an excellent quality eating fish which can reach a considerable size, and its distinctive appearance was already recognised and appreciated throughout its Asian range and in Australia well before it was first cultured on any scale. The market for small, plate-sized whole fish has declined in recent years, and many growers are now focusing on larger fish for filleting because these put on weight more efficiently once they exceed a half-kilogram in weight. Fillets can also be marketed in more diverse ways.

This is a massive and still-expanding segment of the freshwater aquaculture industry, with well-developed technologies for grow-out, specifically prepared diets readily available for every stage of growth, and fairly stable markets with good returns to growers. However, the volume and economics of barramundi production mean that most future increases in production will either be from large-scale corporations entering or buying into the field, or expansion of the many (already large) existing operations.

*▼ Adult sleepy cod with numerous fingerlings – note the passive nature of this fish, which is not struggling even though it is only supported from below.*

Barramundi change sex as they grow, from male to female at around six years old in the wild, but as early as eighteen months in cultured fish. Breeding is in seawater, with juvenile fish moving into rivers as they grow, and the parents may need to be induced into spawning with hormones. Both injection and collecting of the eggs, which may spread through the entire spawning tank, are specialised processes, so many grow-out operations buy in either fertilised eggs or fingerlings rather than produce their own.

As barramundi are cannibalistic they should be graded to prevent loss of fish which are lagging behind in development, so the industry is largely based around intensive culture of one form or another, whether in floating cages in large dams and lakes, raceways or recirculating systems. Some pond culture is still done, and for small-scale domestic production or sport fishing, stocking the fingerlings at up to 100 per hectare in dams with well-established populations of fodder fishes will minimise cannibalism.

**Barcoo grunter** (*Scortum barcoo*) was a recent flash-in-the-pan success story once it was renamed **jade perch** to try to hook the Asian market, where jade is a good luck stone, and the word perch is associated with many better-quality eating fishes. In many ways this species could be regarded as a more tropical version of silver perch, though it is preferred as a live species in Asian markets because its flesh is regarded as sweeter, and it is more tolerant of crowded conditions.

It seems a strange choice given that *Scortum* species as a group are known as leathery grunters for their tough flesh and the noises they emit when taken out of water. However, cultured specimens which are fed regularly develop considerable amounts of fat, regardless of what they are fed on – apparently an adaptation to storing food against the not-infrequent droughts of their natural range. Although food conversion ratios for this fish are markedly better than for silver perch, it should be kept in mind that this is worked out on the basis of total weight gain for amount of food consumed, and that as much as 15 per cent of this can be in the form of fat which would be wasted if the fish was to be sold in filleted form.

*▼ Short-finned eels have the best quality flesh of the Australian species, but the larger marbled eel is more popular in Asia.*

The resulting fish has a rather bloated and not entirely pleasant appearance, and is possibly dropping away in both value and popularity now that it has become familiar. Left to fend for itself in a dam, the fat is only put on when there is abundant live food available, and the flesh becomes firmer. Most jade perch are raised in open ponds in Queensland, and there have been reports of the off-flavour problem in unpurged fish which have been raised in crowded conditions.

**Sooty grunter** (*Hephaestus fuliginosus*) is another fairly large species which has been cultured to a limited extent in the past, but does not appear to be available at the present time. It is regarded as a good angling fish and reaches around 4 kilograms, though 1 kilogram is more common. Like silver perch, it will eat a variety of plant and animal foods including algae, and has the potential to replace silver perch in its tropical range from northern Queensland to the Northern Territory.

It is likely that at least several other grunters (all larger and more tropical species) will also prove suitable for aquaculture, as they are omnivorous and will take pelleted foods. They tolerate fairly crowded conditions with less than optimal water quality, and breeding techniques for the group are already fairly well established. Experiments are already being carried out in hybridising grunters with other species – for example silver perch and Welch's grunter (*Bidyanus welchi*). There are no set stocking rates for all of these fishes, so as a starting point use no more than 250 fingerlings per hectare as has been suggested by Queensland Fisheries for sooty grunter.

**Sleepy cod** (*Oxyeleotris lineolata*, actually a gudgeon) is the current buzz in tropical aquaculture circles, because this fish is showing every sign of being well adapted to intensive aquaculture, and is a high priced species which may justify the expense of running such systems. This is partly because the closely related marbled goby (*O. marmorata*) brings the highest price of any freshwater species in south-east Asia. Sold as a live fish in restaurants, marbled goby is reputed to enhance virility – which may explain the high demand and price to match.

However, marbled goby is already being cultured widely in Asia, and it is likely that once live sleepy cod are exported they will be taken up by the same skilled breeders, while they continue to fetch a premium price for novelty. The same applies to a potential golden form of sleepy cod which is still in development, and has even more exaggerated claims being made for it.

The positive attributes of sleepy cod include its thick, firm flesh and its ability to survive out of water for some time without even struggling, making live transportation simple and relatively inexpensive. Cultured fish can be taught to take pelleted foods from the bottom (though these may be eaten rather sporadically), and will even learn to feed on floating pellets in crowded situations, where competition for food is greater. And the greatest advantage is that sleepy cod don't seem to have the off-flavour problem even when raised in crowded systems, saving on the time and expense of purging.

On the down side, because they are a relatively small fish, they take longer to reach a marketable size. As they rarely reach above a kilogram in weight, they can only be sold as whole fish within Australia, where small fillets are not highly regarded. Being a large-mouthed and territorial nocturnal predator, sleepy cod must also be regularly graded to keep larger fish from eating smaller ones, so they aren't well suited to extensive culture.

This is a largely tropical fish, needing a minimal temperature of 20°C though it can survive 9°C if not otherwise stressed or handled during brief colder periods, and will grow well even at 30°C. It breeds readily if kept uncrowded and provided with pipes or similar shelters, where the male guards and tends the eggs until they hatch. However, the relatively small size (for an aquaculture species) and poor survival rates of fry to date indicate that to produce large numbers of fingerlings it is necessary to keep many pairs of this territorial fish.

**Giant gudgeon** (*Oxyeleotris selheimi*) is a more tropical species which grows even larger, and may take off as a culture species in the wake of sleepy cod. There are also many other future prospects for aquaculture which are still a long way from being turned into any kind of commercial reality. To give just one further example, **jungle perch** (*Kuhlia rupestris*) has attracted recent enthusiasm because it is a handsome species with excellent quality flesh, is favoured by anglers, and is also omnivorous so it may even have a place in polyculture. Unfortunately, there is almost nothing known of its breeding biology, although it is suspected of moving to estuaries to spawn, so it may be possible to produce this species with similar technology to that already established for barramundi.

## *Reading further*

There are numerous regional guides to freshwater fishes in Australia, but the most current overall guide is the *Field guide to the freshwater fishes of Australia* (GR Allen, SH Midgley and M Allen, published 2002 by the Western Australian Museum). *Australian freshwater fishes: biology and management* (JR Merrick and GE Schmida, published by JR Merrick in 1984) is probably the best overall summary of what was known of the biology of many species now being regularly cultured, though there are now diverse more recent papers and articles on particular species available.

*Management of wild and cultured sea bass/barramundi* (edited by JW Copland and DL Grey, published in 1987 by the Australian Centre for International Agricultural Research et al.) remains a primary text on this species, though both research and applied culture have moved on considerably from that time.

For the major species of more southern areas, *Silver perch culture* (edited by SJ Rowland and C Bryant, published 1995 by Austasia Aquaculture for NSW Fisheries) is the definitive work, and much of the information here could also be applied to other, more tropical grunters as well. For the other Murray–Darling species and others *Freshwater fishes of south-eastern Australia* (edited by R McDowall, second edition 1996 by Reed Books) is a good biological guide with much information relevant for aquaculture. *Eel culture* (A Usui, second edition 1991 by Fishing News Books) is worth reading for anyone interested in this group, though most eels in Australia continue to be stocked and harvested from larger lakes and water bodies.

- Raising, marketing and transporting crayfish
- Breeding and stocking rates
- Feeding and purging
- Yabbies
- Marron
- Redclaw
- Spiny crayfish
- Freshwater prawns
- Reading further

*Marron, the largest of the commercially raised freshwater crayfish.*

# 11 Crayfish and prawns

Australia has a rich freshwater crayfish fauna which includes both the largest and the smallest species in the world, with species found from Tasmania to the tropics but with the greatest diversity in the south-east. Many of these grow to a large size, though few grow fast enough to be worth farming, and only three species are presently raised commercially in any numbers. These are the yabby (*Cherax destructor*), the marron (*C. tenuimanus*) and the more tropical redclaw (*C. quadricarinatus*). It is possible that other *Cherax* species will also prove suitable for aquaculture once they are better known, but there is currently no research being done in this direction.

Crayfishes are among the easiest animals to raise in a farm dam as long as they have adequate shelter and good water quality, producing many good feeds over the warmer months. Regular harvesting through trapping is the best way to keep populations from overcrowding themselves, and is useful even in commercial ponds which are to be completely harvested later, giving an indication of how the animals are growing.

## *Raising, marketing and transporting crayfish*

Commercial production is not quite as straightforward as extensive stocking in a farm dam, particularly raising any of the species intensively. Even so, the industry is well entrenched, with an annual value of sales within Australia remaining fairly stable at around $5 million per annum, though this drops somewhat during major droughts. Approximate crop tonnage and prices current for each species as around 2004 are noted under the individual entries, but total production is in the order of 300 tonnes annually. Unlike for the fishes in the previous chapter, most of which have had a history of booms and busts over the last few years, prices for freshwater crayfish have remained relatively stable, so approximate price ranges (as of 2006) have been included for all three farmed species.

There are probably only a few hundred licensed growers of the various crayfish species Australia-wide, and in the last few years many of the smaller operators (particularly in Victoria) appear to be dropping out as increasingly burdensome regulations and charges eat into their profitability. It is likely that the trend to larger farms with more ponds will continue, though multi-farm licenses to harvest from stocked dams on many properties are another and well-established way to supply the markets.

This idea began with yabbies introduced into dams in Western Australia, where they thrived but not in quantities that would justify harvesting and marketing from a single dam or even property. Instead, a more co-operative approach was adopted, with individual farmers trapping whatever they could at times when other work wasn't paramount, and delivering these to a centralised point where they were purged and graded before sale. This system worked so well that Western Australia has developed a substantial enterprise from what could otherwise have just remained a cottage industry, and is the largest producer of yabbies in the country.

Grading is an essential part of value-adding for all crayfish species, especially for the relatively lucrative supply of restaurants, which like a regular supply of clean-shelled, purged, undamaged, live animals with matching claws, and all within a pre-agreed size range – usually large. Even a grower with just two or three ponds in production should be able to make a significant part of a living from freshwater crayfish if the prime specimens can all be sold direct to an accessible restaurant.

Much has been made of the necessity to be able to supply quality product year-round, but many of the better-paying restaurants are in seasonal tourist areas, particularly in coastal areas. Here, demand is most concentrated in the warmer parts of the growing season, falling away during cooler times of the year when crayfish may not be feeding or growing.

Freshwater crayfish are able to survive some time out of water, so they can be freighted around by purging at 15°C to 18°C for two or three days, then cooling them off and packing in moist (but *not* wet) materials such as wood shavings in a sealed foam box. Although virtually all successful crayfish growers raise the animals in outdoor ponds, purging is done in recirculating systems or, for smaller operations, in lidded crates with clean water trickling down from above. Even when just bringing home a bucketful for a meal, rinse them off in cool water before draining all surplus liquid, as crowded crayfish in a small volume of water will suffocate, and even a small amount of polluted water at the bottom will damage their gills if it contacts them.

## *Breeding and stocking rates*

Crayfish grow in jumps, rather than in an even progression, by shedding their old shell once the soft parts of the animal within are crowding it. This is called moulting, and the soft-shelled animal which emerges must have a shelter to retreat to or it may be cannibalised by others of its kind. The shell hardens within a few days, the animal turning into a handsome looking, clean-shelled crayfish, but the meat within is sparse compared to its overall size until it has had time to put on some additional flesh.

Mating takes place when a mature female moults, and the male will usually stay with the female to protect her while her shell hardens. The sexes are easily separated by looking at the bases of their legs, where males have protrusions on the last pair, while females have a pair of rounded oviduct openings on the third pair from the back. The female later lays her eggs so that they attach to the little fins (swimmerets) under her tail, which she keeps curled up to protect them as they develop into miniature crayfish. Females carrying eggs are referred to as berried.

Unrestricted breeding of any crayfish species in a pond brings growth effectively to an end as the animals divert energy into reproduction. A secondary effect is crowding, reducing the amount of food available, and adversely affecting water quality if extra food is added to compensate. In turn, worsening water quality will put most crayfish off their feed, which is why production can virtually come to a halt in crowded farm dams during drought years.

There are various ways to prevent overbreeding, including stocking only one sex in a pond – a time-consuming process which is not often

▲ *Yabby gonads on the base of the last pair of legs on the male at left, and on the base of the third pair on the female at right.*

done on a commercial scale, requiring initial sorting then fencing between ponds to stop animals wandering over to join a pond of the other sex. Another method is to stock only one size of young crayfish, raised in a nursery pond until they are large enough to all mature and be harvested after a single season's growth. As both yabbies and redclaw may mature at 20 to 30 grams body weight, especially in crowded conditions, even this method may just result in a pond of stunted animals saleable only as bait.

The easiest population control method to implement in farm dams is adding a suitable predatory fish that will feed only on the youngest animals, though growth rates of breeding adults will still be reduced to some degree. Silver perch are good for this, but more carnivorous species such as golden perch or catfish should only be added while small, and should be removed before they grow large enough to feed on smaller adult yabbies.

An adequate amount of shelter in the form of brush and irregular branches, shadecloth, lengths of pipe, and similar hollows will reduce losses to cormorants, and discourages yabbies in particular from burrowing. Tyres have also been widely used, stacked in rows at a slant against each other, but their use is being discouraged on a precautionary basis in some areas, despite lack of any evidence that they affect water quality adversely. Steep pond sides and vegetation along the shallower edges also make fishing harder for wading predatory birds such as herons.

▲ *Female crayfish carrying eggs under their tails are referred to as berried; the tail is normally kept curled up at this stage to protect the developing young which are released as miniature crayfish.*

Yields for the three species are roughly comparable, though redclaw have the advantage that in areas where they grow all year around, more harvests can be made from the same ponds over any given time span. Crops are sometimes surprisingly variable, and even now no one seems to know why one pond on a single property would produce only the equivalent of a few hundred kilograms per hectare in the course of a year, while another seemingly identical pond yields the equivalent of two tonnes and more. Exceptional annual yields of up to 4 tonnes per hectare for yabbies and marron have been recorded occasionally, and up to 5 tonnes for redclaw.

However, at the stocking rates required to achieve such results (up to 8 or even 10 animals per square metre), management must be impeccable at all times, from feeding and grading to keeping track of water quality. Having the reserve capacity to change at least part of the water is desirable, and there must be abundant shelter to minimise the dramatic differences in size seen in crowded ponds. The longer a heavily stocked pond is left unharvested, the more size variation is likely to become significant, with animals up to four times the size of their smaller siblings dominating the pond and reducing the average growth rate.

◄ *Tyres have long been used as an inexpensive form of shelter; here they are stacked out of the way during pond cleaning at Crayhaven.*

Lower stocking rates of around two to three animals per square metre give better individual growth rates on average, and also less variation in size overall. Given the need for increased management and a greater, hands-on familiarity with crayfish biology and behaviour, newcomers to the field should probably not attempt to stock more than two animals per square metre while they experiment with what can potentially be a very delicate crop when pushed beyond sensible limits.

## Feeding and purging

Crayfishes are largely vegetarian, even though the favoured method of catching them is with a piece of (fresh) meat on a string, dip-netting the animals out once they take the bait. They will eat a range of vegetable foods and the microscopic organisms associated with decay, including rotting grass or hay and algae. Crayfishes will also catch smaller water animals including each other when food is in short supply, and even slow-moving fishes at night.

Pelleted foods specifically made for crayfish are readily available for commercial growers, and appear to be pretty much interchangeable between species in terms of resulting growth rates. Farmers seeking to cut costs of feeding in extensive culture often use cheap feeds such as lupin seed successfully, though bulk fillers such as chicken pellets are more likely to be ignored and cause water quality problems. Carrots are another popular and inexpensive feed, and are devoured particularly eagerly by yabbies. Varying feeds or adding them in different combinations may keep the animals feeding more vigorously, as some growers suggest that boredom can be a factor in crayfish going off their feed.

Crayfish are disease- and problem-free compared to fish, and few of their problems will transfer to other individuals except in the most putrid pond conditions. In fact, most problems seen are the result of poor management and stress, and improved water quality and feeding will usually see any signs of damage disappear after the next moult. Even white-tail disease (*Thelohania*) doesn't seem to be infectious in uncrowded ponds, and the few animals which gradually waste away from this are usually the first to be picked off by predators so the problem may solve itself. Regular draining and removal of accumulating sludge in commercial ponds will also help minimise such problems.

Cosmetic problems are more severe from the marketing point of view, especially where crayfish have dirty shells with algae or various micro-organisms forming an unattractive layer over parts of the shell. Some are attached to the shell, while the stubby, brown worm-like animals known as temnocephalids move around on the shell; although these don't harm the crayfish and they drop off during cooking, they certainly won't impress most customers if present in large enough numbers to be conspicuous! Most shell growths will disappear after the next moult in cleaner but turbid waters, or they can be cleaned up with a 10-minute bath in salted water at least a third the concentration of seawater (all three farmed *Cherax* species will tolerate this level of salinity indefinitely, and yabbies can even survive straight seawater for days).

## Yabbies

Yabbies were the first of the native freshwater crayfishes to attract interest from aquaculturists, and they thrive in shallower, warmer areas of ponds and dams so they could be raised on a considerably larger scale for personal use, if not commercially, by stocking many otherwise unused farm dams. Most farmers are reluctant to do this, however, because yabbies burrow into the clay walls, particularly when water levels are falling, or where the water is clear so they are exposed to predators.

Found naturally across a large part of inland eastern Australia down to southern Victoria, these are the smallest of the three farmed crayfish – even so, larger specimens can reach a respectable length of around 25 centimetres and a weight of 300 grams, and I recall several caught in the 1967 drought year which were over 30 centimetres when stretched out. Yabbies are very variable in colouration, ranging from murky cream through greens and blues to near black, though the colour of many wild populations is affected by the waters they live in, and may change after their first moult under different conditions. Some differences are genetic, and the blunt-faced, smaller-clawed south-eastern Victorian populations are regarded as *C. destructor* subspecies *albescens* by some scientists, and a full species *C. albescens* by others.

Yabbies have also been introduced to Western Australia and onto Kangaroo Island in South Australia, where (along with marron) they have taken off with a vengeance and completely altered the ecology of freshwater streams on that island. This species is declared noxious in Tasmania to prevent its introduction there. In Victoria, many smaller-scale yabby growers have dropped out of the industry after the implementation of ridiculous laws which require expensive and unnecessary on-farm inspections of live product, completely out of keeping with all other farm livestock which is only inspected at an abattoir at the slaughter stage.

Yabbies will tolerate a wider range of temperatures than the other farmed species, as well as survive droughts by burrowing deeply. Their extreme adaptation to arid conditions includes the habit of walking out of waters which are deteriorating in water quality or evaporating fast, and looking for a home elsewhere. The claws on this species are large in proportion to the body size, which is regarded as a disadvantage by people who think only in terms of proportion of tail meat, but many others (including myself) regard the meat of the claws as particularly fine in quality.

This is the lowest valued freshwater crayfish, though its flesh is sweeter and less grainy than that of the marron, which commands a premium price mainly because of its much greater size, and the longer time it takes to produce an unusually large specimen. Commercial production Australia-wide of the yabby is around 120 tonnes, though this fluctuates with drought years, and it is probable that many more are being harvested from the wild or for private use. Prices in 2005 varied considerably with size and quality, from around $8 to $13 per kilogram, though well-matched groups of clean, undamaged, larger yabbies can fetch a premium $15 to $18 if sold direct to restaurants.

For commercial production in a pond with only one size stocked for a later drain harvest, a density of two yabbies per square metre should reach a harvestable size within a single growing season. If they are to be grown to a larger size for a better price, half this rate is better. In a polyculture situation lower numbers can be used initially, but as breeding is likely to be continuous and ongoing for many years just adding a few mature adults may be all that is needed.

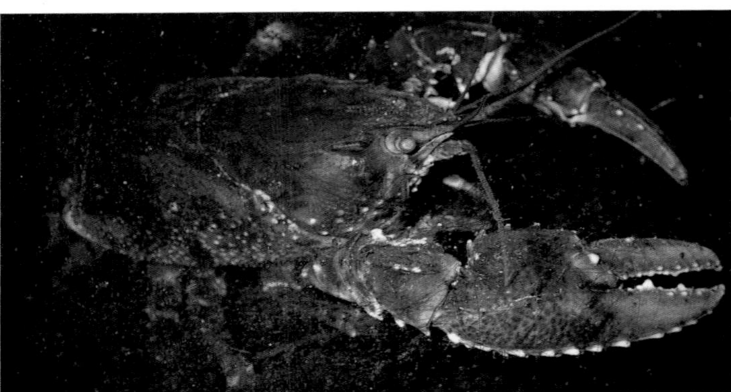

◀ *Many spiny crayfish grow to a large size but too slowly for commercial production – this is* Euastacus bispinosus.

▶ *A large* Macrobrachium rosenbergi *showing the characteristic claws of the male. This freshwater prawn is cultured in south-east Asia, and is also native to northern Australia.*

## Marron

◀ *Yabbies, the first native crayfish to be commercially farmed successfully.*

Marron are a larger cousin of the yabby, originally from the south-western corner of Western Australia but also very successfully naturalised on Kangaroo Island. Even though they have much more specific environmental requirements than the yabby (low dissolved oxygen and temperatures outside the range of 15–25°C stress them so that they stop growing) they have been prohibited from many places including all of Victoria. The attraction of this species is its large size, up to 2.5 kilograms, with smaller claws and therefore a greater proportion of tail meat for its size than the yabby. A nice specimen of this species makes a meal in itself, and can fetch a premium of up to $24 per kilogram.

Marron farming is mostly confined to its native range and also Kangaroo Island, with total production around 80 tonnes per annum. Virtually all successful farms are in popular, coastal tourist areas where premium prices are more readily achieved, whether from farm gate sales or direct to restaurants specialising in local produce. This is important, because marron don't grow any faster than yabbies so it takes a lot longer to raise a premium-sized animal.

As a polyculture animal, marron is only really useful in mild, coastal climates in temperate areas – it can't compete with redclaw in consistently warm climates, or yabbies anywhere that all-round hardiness is important. One advantage of this species is that marron don't burrow, so they need shelter to be provided in open dams if unprotected by netting. As this is potentially a large animal and the largest marron fetch the best prices, stocking at no more than one per square metre is close to ideal, or twice this if the faster-growing ones can be removed earlier. Individual animals may grow at very different rates so that some

▼ *Redclaw are handsome beasts best suited to tropical and subtropical conditions.*

may be four times the weight of their smallest siblings, though this may possibly be overcome in the long term by selective breeding.

## *Redclaw*

Redclaw are a tropical to subtropical species which aren't commercially viable south of Queensland because temperatures are too low for sustained growth for much of the year. Their native range extends from the Northern Territory around the Queensland side of the Gulf of Carpenteria, and they are now farmed along the eastern coastal areas of Queensland as well.

An attractive animal with a distinctive red-bordered claw, redclaw reach around twice the weight of a yabby and their flesh is of comparable quality. They are also comparably hardy to yabbies, but tend to stop growing once temperatures fall much below 20°C, and are apparently less tolerant of extreme turbidity. Redclaw production is much higher today than it was in the 1990s, and is probably similar nationally to that of marron, though prices are lower at anywhere from $12 to $20 per kilogram with larger animals (as usual) fetching a premium.

As this is a tropical species, more crops can be squeezed into a year, harvesting up to twice, though growing periods of 8–9 months are more usual. The downside is that, in tropical regions, mixed-sex stockings can continue to breed for much of the year, spending energy on reproduction rather than growth. In more southern parts of Queensland, the water temperature is suitable for breeding only for a relatively short time and so, apart from around three warmer months, the animals put their efforts into growth for most of the year.

Harvesting this species is relatively easy as they are strongly attracted to clean water flowing into their ponds, and move upstream into the current so they can be collected in unbaited flow traps. These can be units mounted outside the pond with water running down a ramp at the side, so the marching hordes reach the top of the ramp and fall into a smooth-sided container from which there is no escape. Submerged flow traps are more versatile, as by using different grades of mesh they can be set to collect juveniles only, or allow juveniles to escape while retaining the saleable-sized animals.

## *Spiny crayfish*

There are many species of spiny crayfish found in south-eastern Australia, mostly in the genus *Euastacus*, and quite a few of them grow to a substantial size. These include the Murray cray (*E. armatus*) which reaches 3 kilograms and around 45 centimetres in length, while the closely related giant Tasmanian crayfish (*Astacopsis gouldi*) is the largest freshwater crayfish in the world at up to 4 kilograms and almost 60 centimetres in length.

Though these large and striking animals are tempting prospects for aquaculture, they are mostly from cool, moving waters where there is plenty of room to spread out, and grow so slowly that there is not the slightest prospect of making a profit from raising them, except perhaps in a small way as spectacular ornamentals for aquaria. They will grow well enough in deeper farm dams and are compatible with trout, but as an edible animal have the further disadvantages of being so spiny that they can cut hands if carelessly shelled after cooking, and they yield little meat relative to total weight.

## *Freshwater prawns*

There are several large-growing freshwater prawns in Australia, all in the genus *Macrobrachium*, with distinctive long pincer arms on the males. They are good eating as well as good looking, and are highly regarded in south-east Asia where the species *M. rosenbergi* is cultured. This species is also found in tropical northern Australia, and is occasionally sold through the aquarium trade. The much smaller, temperate species *M. australiense* also reaches an edible size but is mostly only cultured as a bait animal.

Despite well-established culture techniques, *M. rosenbergi* has not been farmed much in Australia except experimentally, partly because the various larval stages (of which there are many) all have specific live food requirements which are labour intensive to provide. The larger and more readily saleable males are also territorial, which makes keeping this species intensively difficult and, with the existing competition from wild harvest of marine prawns, it is unlikely that anything much will be done with freshwater prawns in this country in the near future. However, prawn lovers in the tropics could certainly produce small crops for their own use in outdoor ponds, treating them much as redclaw but allowing more space.

## *Reading further*

*The commercial yabby farmer* (R McCormack, published 2005 by RBM Aquaculture) is the best and most complete practical manual on the subject, covering a wide range of issues encountered by growers and the author who has been raising crayfish for a very long time in semi-intensive ponds. Anyone seriously looking at commercial farming will find it also worth reading *The Australian yabby farmer* (J Mosig, 2nd edition, 2000 by Landlinks Press). The only general biology of the farmed species is *The yabby, marron and redclaw: production and marketing* (JR Merrick and CN Lambert, published 1991 by JR Merrick Publications).

For a complete overview of freshwater prawn farming in Asia see *Freshwater prawns* (edited by EG Silas, published 1992 by Kerala Agricultural University, Kochi, India) and, better still if you can find it, *Freshwater prawn farming: the farming of* Macrobrachium rosenbergii (MB New and WC Valenti, published 2000 by Blackwell Science, Oxford).

*Almost every part of the spectacular sacred lotus* (Nelumbo nucifera) *is edible.*

- Growing conditions
- Weed warning
- Edible plants
- Plants in polyculture
- Hydroponics
- Water quality plants
- Other commercial plants
- Reading further

# 12 Plants

Aquatic plants have generally been perceived as weeds in mainstream aquaculture, things which get in the way of harvesting or reduce available oxygen levels at night in warm weather. This is largely a function of ignorance and poor selection, or no selection at all when genuinely weedy species (of which there are plenty) are introduced by waterbirds and grow out of control.

Some fish and crayfish growers have experimented with waterplants over the years, and there is an increasing though slow acceptance that they can be of value if chosen to be compatible with the animals being farmed, as an inexpensive type of shelter providing shading and protection from predatory herons or cormorants, as food for herbivorous species, or to strip nutrients from waste water. Many are also saleable in their own right, whether as ornamentals, stock feeds or even as human food.

## *Growing conditions*

Before looking at any of these aspects, it is useful to loosely explain the ecological groups of waterplants, and the way they grow. True aquatics grow in water, whether deep or shallow; some remain submerged at all times, others develop floating leaves, and still others (often referred to as macrophytes) have emergent leaves or stems. Most true aquatics will tolerate some drying out once they are well established, as water levels fall through evaporation in a dry summer, though their tops may die down for a dormant period.

Plants for waterlogged soils will usually tolerate flooding for a short time, or even weeks in some cases, but prefer to grow with their roots in sodden soils where there is little or no oxygen. Perhaps 'prefer' is too strong a word – they are able to survive here because they can shift oxygen down to their roots, giving them the edge in this habitat over water's edge species which would drown if their roots were flooded for more than a few days.

The shoreline and small hummocks which rise above the waterlogged zone are where water's edge plants grow. These are species needing moist to wet soil with air spaces present, so some oxygen is available to their roots even if there isn't a lot of it. A final category is the plants of seasonal wetlands, which may fit into any of the categories above during the wet season – winter and spring in temperate zones, or during monsoon rains in the tropics. These need a drying-out period to ripen fruits, tubers or seeds, and include many of the most productive edible plants including Chinese water chestnut, rice and other water-loving grains.

Plants from deeper waters rarely need much more than a 5-centimetre deep layer of soil to take root in, as they meet much of their nutrient needs direct from the water. In shallower waters a soil layer 10 centimetres deep is adequate, and this can be up to 15–20 centimetres deep in the waterlogged zone, so plants with exposed tops can anchor themselves solidly against wind and wave action. Few plants will grow much deeper than a metre even in the clearest waters, so laying topsoil at greater depths is unnecessary unless you have some other reason for using it there.

Soils for most aquatic and water's edge species should be on the acid side, with little or no trace of lime or any other alkaline substance. Soil pH is more important than water pH, and if the soil is suitable it doesn't matter so much if the water is a little too acid or alkaline – the plants will be able to obtain most of their requirements through their roots. Most topsoils are suitable, as long as they are neither too sandy nor heavy clay.

Fertilisation is best done with just blood-and-bone, a nitrogen-rich abattoir by-product which makes an excellent slow-release fertiliser when submerged. Synthetic fertilisers are based around inorganic salts, which many aquatics dislike, and should be avoided. These aren't formulated for aquatic plants in any case, and as they are highly soluble they leach readily into the water, where they can trigger algal blooms and similar problems. Some fertiliser tablets specifically made for waterlilies and similar species are available, but these are by far the least economical option, despite ease of use.

## *Weed warning*

Before looking at some uses of plants, be warned that many aquatic plants are weedy, and some of them among the most serious weeds in the world. With the wide range of potentially weedy species already available in Australia, and virtually every possible climate type for them to grow in, it should be compulsory for anyone introducing any plant from outside the country (or even from outside its native range *within* the country) to research the species in detail before planting.

There are already extensive lists of wetland and aquatic plants which are illegal to bring into this country, but they are inadequate as there are perhaps hundreds of others with at least some weed potential which remain allowable imports. A better system would be to implement a strictly itemised list of allowable imports as has been done for fishes, so that any new potential entry must be thoroughly researched and cleared of weed potential through a tribunal before it can be added to the list.

Given that there are many more plant species in the world than there are freshwater fishes, this approach would seem to be commonsense, and as far as I can see the only reason it hasn't been implemented is because gardeners are possibly perceived by politicians as a significant voting bloc. If this is the case, now would be an excellent time to apply more restrictive rules to the range of plants allowed into the country, as present garden fashions lean more to water features with spinning ceramic balls and statuary than to anything green that might actually grow!

If there is a suspected weed potential for any species of plant, experiment carefully with it in an enclosed container before planting out, especially if runoff from your aquaculture can carry pieces of the plant or (worse still) its seed into wetlands downstream.

## *Edible plants*

Many aquatic plants have been used as food through history and pre-history, and the best of these have been cultivated and selected to provide some of the most productive agricultural crops. Aquatic species are particularly important in south-east Asia, which was a probable centre for domestication and selection of many significant species from arrowheads to sacred lotus and trapa nuts. Two of the earliest crops brought into cultivation were probably domesticated here. Taro is a major staple through the warmer Pacific and has been spread worldwide, while rice is the single most important plant crop in the world.

Only a tiny handful of edible plants have been tested in conjunction with aquaculture. Almost invariably, those which fetch very high prices, or those which are astronomically productive draw the most attention. Chinese water chestnut (*Eleocharis dulcis*) is a the classic example of a very productive species which can potentially produce over 100 tonnes of crisp, starchy but sweetish tubers per hectare over a 7-month growing season, though yields up to 50 tonnes are more realistic.

*▲ The large tubers of taro, one of the earliest agricultural crops and still a staple in many tropical countries.*

Promoted widely as a miracle monoculture (though in Asia it is grown in rotation with arrowheads, sacred lotus and rice) in the 1990s, it is now rarely available, and fresh product is almost never seen, even in the Asian markets in Australia which were supposed to be such a pushover for this plant. Some poorly thought-out assumptions underlay the promotion of this easily grown plant, one being that as a novelty it would sell at fanciful boutique prices which never eventuated.

▶ *Plants of Chinese water chestnut, one of the most prolific tuber-producing plants known.*

▼ *A native water ribbon* (Triglochin procera) *showing sweet, crisp, peanut-sized tubers on 12-month-old seedlings.*

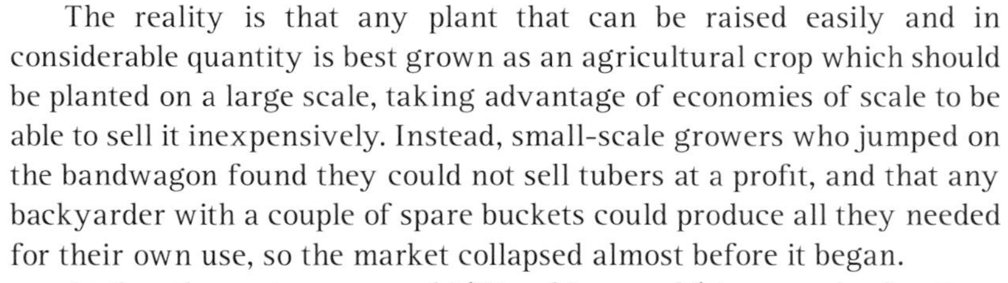

The reality is that any plant that can be raised easily and in considerable quantity is best grown as an agricultural crop which should be planted on a large scale, taking advantage of economies of scale to be able to sell it inexpensively. Instead, small-scale growers who jumped on the bandwagon found they could not sell tubers at a profit, and that any backyarder with a couple of spare buckets could produce all they needed for their own use, so the market collapsed almost before it began.

At the other extreme, wasabi (*Wasabia wasabi*) is a genuine boutique crop which fetches very high prices (up to around A$150 per kilogram) in Japan, and is being experimented with as a water treatment plant for trout farms. This is the condiment plant used to flavour the hot, spicy and green (usually dyed) paste used with sushi. Sales of this species to date seem to be confined to Australian growers keen to get into this theoretically lucrative market, and I am not aware of any significant exports apart from for trial purposes.

Reading the early accounts of the first trials in this country, it is clear that the proponents had not done much research on wasabi. It is only the highest grades of stems which sell fresh at premium prices to a relatively small gourmet market, and these take many years to mature so the trials which aimed to produce a saleable crop in the shortest time were pointless. There is even the suggestion that older, virused plants fetch the best prices.

Most wasabi is sold at much lower prices for making cheap paste, dyed a vivid green to imitate the colour of the premium product, and adulterated with horseradish to bring the price down as this is very similar in taste. The reason for the adulteration is not a worldwide shortage of low-grade wasabi (though this may come to pass as Japan's cities continue to expand), but that horseradish is cheaper because it is much easier to grow. The only reason that wasabi is needed at all is so that its prestigious name can be used on the paste!

There are around 300 species of aquatic and water's edge plants worldwide which have been grown or used for food, with literally tens of thousands of cultivars and varieties, and a considerable number of these are available in Australia. With a complete spectrum of Australian climates from tropical to cool temperate, there is certainly potential to incorporate many plants into mainstream aquaculture. This may be in separate ponds which can be used to treat waste water while generating a useful by-product in their own right. Others need to be grown in ponds which can be drained separately for maturing and harvesting their crop, or in waste water that can be used to irrigate entire paddy fields for larger-scale horticulture.

Some species may prove to be more directly useful in water treatment, for example watercress (*Nasturtium officinale*), which develops soft, trailing roots, an ideal attachment site for micro-organisms that utilise and detoxify fish wastes – and it may also take up ammonium directly. A salad plant with refreshingly peppery foliage, it grows well only in permanently cool, flowing waters and could combine water treatment with a saleable and easily harvested vegetable.

In shallower ponds, nardoos (particularly *Marsilea mutica*) can provide a dense stand of upward trailing stems that shelter young crayfish from predatory birds, yet won't grow more than around a half-metre deep so they won't interfere with a drain harvest. The exposed tops of these water ferns then break down to fuel plankton blooms later, and the edible sporocarps can be harvested as they separate from the drought-tolerant roots, which will regrow new leaves within a few weeks when the pond is refilled. Common nardoo (*M. drummondii*) is best known as the bush food that explorers Burke and Wills were completely dependent on when they died of hunger, but the Aborigines who taught them its use were more skilled in its preparation, and only included it as part of their diet, as it is not a complete food in its own right.

Other fast-growing natives such as swamp lily (*Ottelia ovalifolia*) have been used successfully as a forage and shelter combined for yabbies, and are soft-leaved so they don't get in the way of farming activities, regrowing from seed once a drained pond is refilled. Although this species is not known to be eaten by humans, its close relative *Blyxa aubert* is still faster-growing in tropical conditions, forming large plants within a few weeks of rain. This can be harvested as human food as well, and is highly regarded in parts of Asia. It is also native to most of tropical Australia where it is probably an important habitat plant.

## *Plants in polyculture*

Apart from their potential as crops in their own right, many plants can be fitted into polyculture systems, becoming not just a supplementary source of food but in some cases a major crop, with the cultured animals being the incidental benefit. Allowance needs to be made at the planning stage for specific areas to grow these in, so it is useful to have a clear idea of which species are going to be used, rather than improvising after construction is complete.

Planting shelves, plateaus and other areas of uniform depth are the most straightforward way of confining plants to a pre-selected area, offering them ideal conditions and depths only where you want them to grow. Thus, a plant such as one of the water ribbons (*Triglochin*, with many large-tubered, edible species in different parts of Australia) which will grow best in, say, 30 centimetres depth should have a suitable planting area dug around 40 centimetres below the future maximum level of the pond. This allows for a 10-centimetre layer of better quality soil to cover the shelf.

Any fertiliser to be used should be spread on the shelf before the soil is laid over it. This minimises nutrient leaching into the water, while leaving the fertiliser readily available for plant roots as they grow down. For plants such as rice which need a drying-out period to form seed, the planting shelf should be planned so it will be exposed by evaporation towards the end of the growing season. This may be a few months after the wet in the tropics, or mid to late autumn in the south.

## *Hydroponics*

Hydroponic culture of plants using aquaculture waste water has attracted considerable attention over many years, using terrestrial plants (particularly lettuce) to mop up some of the wastes produced by intensively cultured fish. It seems a good idea in theory, but the resulting systems are disappointing in both environmental and economical terms when analysed in any detail.

Most commercial hydroponic systems use long runs of pipes or troughs filled with a light, inert medium such as vermiculite or perlite, through which a balanced nutrient solution flows, with variations in nutrient level adjusted by a computer. The soil-less medium is supposed to reduce pest problems (though toxic sprays may be necessary if any disease or insect pest appears), and growth is fast as the entire system is often set up in a greenhouse.

The aquaculture version of hydroponics uses waste water from intensive ponds which must often be stored in a holding tank until needed, with aeration to prevent anaerobic decay releasing methane, hydrogen sulphide and other unwholesome gases. These wastes don't constitute a balanced liquid fertiliser for the plants, but must have additional nutrients added by the same type of computerised system. Once water has passed through, it must usually be disposed of by irrigation, as the excess of unnecessary nutrients makes it unsafe in fish ponds, and if any spraying has been done to control insects or other pests then it will be positively toxic.

The problems with this unnecessary combination of two high-tech systems are manyfold. Of the major ones, the additional and considerable

▶ *Kan kon, one of the fastest growing and most productive green vegetables of the tropics, thrives in nutrient-rich waters, even in southern summers.*

▲ *Watercress grows abundantly in cold, flowing waters, and is an ideal water quality plant for trout farms. The mildly peppery leaves and stems are a popular and nutrient-rich salad vegetable.*

expense of a hydroponic system is an added financial burden on setting up the intensive system which is presumably meant to be the primary source of income. This would be enough to make all but the best-funded ventures non-viable. Secondly, it would be far less expensive to simply irrigate crops directly in the ground, rather than having to erect kilometres of pipe, additional greenhouses, high-tech storage tanks, and then have to computerise the lot.

Outdoor crops would not grow as fast, but a far greater diversity of seasonal vegetables could be produced from the same waste water, simply by watering as required. Outdoor crops are also less prone to various fungal and insect problems which can build up rapidly in a greenhouse, so toxic chemical input is less likely to be needed, and if grown by a skilled organic grower should not be needed at all for a market garden scale of operation.

At another level, it is hard to see why there is such enthusiasm for this unwieldy approach to dealing with waste aquaculture water, as terrestrial plants mostly take up nitrogen in the form of relatively non-toxic nitrates. The biological process of nitrification (see chapter 2) turns ammonium into nitrate,

▶ Nymphaea *'Mrs Pring', an ornamental tropical waterlily hybrid.*

lowers pH and also uses up oxygen. By contrast, many aquatic plants are known to take up the much more toxic ammonia directly (in the form of ammonium) in preference to nitrates, cutting short the breakdown cycle via equally toxic nitrites.

If they only have nitrate available, aquatic plants will use it, but it must first be converted back to ammonium through nitrate reduction, which takes energy. Some aquatic plants which are known to use this cycle are productive edible species, for example the nutrient-rich water egg (*Wolffia arrhiza*) which can produce ten tonnes *dry weight* per year just through regular harvesting from small ponds – and 20 per cent of this is protein,

with 44 per cent carbohydrate. Further research will undoubtedly reveal many other edible aquatic plants which use ammonium directly, skirting the nitrogen cycle entirely, and at the same time producing a saleable crop without the elaborate and expensive equipment needed to produce a hydroponic lettuce.

## *Water quality plants*

Water treatment wetlands have already been discussed in chapter 3, and all that needs to be looked at here is what plants actually do to improve water quality as it passes through a wetland. Although this varies between species, the changes may include trapping of sediment and removal of some heavy metals, buffering of pH and stripping of useful nutrients in much larger quantities than a comparably-sized terrestrial plant could do.

This is called luxury uptake, where essentials for long-term growth such as potassium and phosphorus may be stored within the plant in considerable quantities, sometimes five or even ten times greater than its current needs. If a shortfall of these nutrients starts to limit the growth of the plant at some other time, they are released gradually and used so that normal growth can continue without interruption.

The beauty of a treatment wetland is that even if there never is a shortfall, the plants will continue to stockpile more and more surplus as they grow larger. In other words, the plants are removing far more of the potential pollutants than they actually need, and some of this surplus can be disposed of from time to time by removing a certain amount of the living plant matter from the wetland. Composted or used for other purposes elsewhere, this becomes a true recirculating and cleaning system for water.

A heavily planted and well-established wetland which is large enough to deal with the waste water volume from an aquaculture farm should not only improve water quality, but if bird numbers are kept low it may also clean water so well that it can be re-used in the aquaculture ponds themselves. A similar result may be achieved in a smaller wetland by recycling the water through with pumps several times, though this will limit the capacity of the system to take additional waste water. This cannot be kept up indefinitely, of course, and a certain percentage of fresh water must be introduced as well in crowded systems, but it can make a substantial reduction in water costs or even make a venture viable when there is not enough water right available to make it feasible otherwise.

## *Other commercial plants*

Some waterplants are raised primarily for ornament rather than food, as water gardening has been a rapidly growing segment of the gardening industry for the past two decades. There is some overlap of course – for example the spectacular lotuses (*Nelumbo* species and hybrids), every part of which is edible. However, potential growth in this field is increasingly limited as many new water garden nurseries have appeared over the past few years, and some larger wholesalers are now specialising in producing self-contained aquatics for smaller nurseries without ponds to display them in.

A more specialised ornamental area is the production of aquarium plants, though this is mainly catered for by a single large nursery in southern Queensland supplying most aquarium shops country-wide. The possibilities here are still further limited by the small percentage of aquarists who have any interest in plants rather than fish, and the limited range of species which will grow in low light conditions.

Equally specialised are the indigenous wetland nurseries which raise plants on a large scale for water treatment, planting constructed wetlands, landscaping and to a lesser degree habitat restoration. Demand for these plants is much greater than for most ornamental purposes, and given that the largest nurseries of this kind tend to congregate in major cities, and many species of native aquatics are found in other places altogether, there is probably still considerable scope for smaller nurseries specialising in regional wetland flora.

Other aquatics have more specialised uses, many of them medicinal, including some which can only be imported in dry form but are much more active when used fresh. These include gotu kola (*Centella asiatica*), water hyssop (*Bacopa monniera*) and dozens of lesser-known species used by Chinese herbalists. Still less well-known are the water's edge species which can be used as unusual cut flowers for the floristry industry, the most spectacular of which are the cold-tolerant North American pitcher plants (*Sarracenia*).

## *Reading further*

*Edible water gardens: growing waterplants for food and profit* (N Romanowski, published 2006 by Hyland House) is the one and only worldwide guide to edible aquatic and water's edge plants, including their cultivation and with extensive references for the various species. For a general guide to planting constructed wetlands using Australian species see my earlier book *Planting wetlands and dams: a practical guide to wetland design, construction and propagation* (published 1998 by UNSW Press), and for ornamental plants within Australia, see my earlier *Water garden plants and animals: the complete guide for all Australia* (published 2000 by UNSW Press).

# Appendix A
# Some sources of aquaculture animals and plants

The suppliers included here offer starting stock (usually fingerlings) of diverse edible fishes and crayfish from juveniles to broodstock, or edible aquatic and water's edge plants and indigenous plants for water treatment wetlands. These lists are far from complete, as many agricultural suppliers in country towns also offer a regular delivery service from commercial hatcheries. Note also that most hatcheries will add to their range whenever possible, so it is worth checking what new species are available from any one source before looking further afield for them.

**Alphatech Aquaculture**, PO Box 98, Old Noarlunga, SA, 5168. (08) 8386 3555. Barramundi, jade perch.

**Ausyfish**, PO Box 324, Childers, Qld 4660. (07) 4126 2226. Barramundi, bass, catfish, jade perch, golden perch, silver perch, sleepy cod. <http://www.ausyfish.com>

**Ballarat Trout Hatchery**, Gillies St, Ballarat, Vic, 3350. (03) 5334 1220. Brown and rainbow trout.

**Blue Water Barramundi**, Mourilyan Harbour Rd, Mourilyan, Qld 4858. (07) 4063 2455. Barramundi.

**Buxton Trout Farm**, 2118 Maroondah Hwy, Buxton, Vic 3711. (03) 5774 7370. Brown and rainbow trout.

**Crayhaven Aquaculture Farm**, 6408 Pacific Hwy, North Arm Cove, NSW 2324. (02) 4997 3002. Bass, catfish, golden perch, silver perch, yabbies.

**Darwin Aquaculture Centre**, Channel Island, Darwin, NT 0800. (08) 8924 4259. Barramundi.

**Dragonfly Aquatics**, RMB AB 366, Colac, Vic, 3249. Edible and water treatment plants, smaller native fishes. (03) 5236 6320.

**Freshwater Aquaculture**, Bushgrove Retreat Rd, Uralla, NSW 2358. (02) 6778 7140. Jade perch, Murray cod.

**Gladstone Water Board Hatchery**, corner of Lord St and Glenlyon St, Gladstone, Qld 4680. (07) 4972 9548. Barramundi.

**Glencoe Fish Hatchery**, 2333 Loddon River Rd, Appin South via Boort, Vic 3578. (03) 5457 8218. Murray cod, yabbies.

**Glenwaters Native Fish**, Break-O-Day Rd, Glenburn, Vic 3717. (03) 5797 8384. Bass, catfish, golden perch, Murray cod, silver perch, yabbies.

**Granite Belt Fish Hatcheries**, Donges Rd, Severnlea, Qld 4352. (07) 4683 5283. Golden perch, Murray cod, silver perch.

**Hanwood Fish Hatchery**, 323 Redgate Rd, Murgon, Qld 4605. (07) 4168 1558. Barramundi, bass, catfish, golden perch, jade perch, silver perch.

**Huon Aquaculture**, PO Box 1, Dover, Tas 7117. (03) 6295 8111. Rainbow trout. <http://www.huonaqua.com.au>

**McHale Fisheries**, 53 Hodgens Rd, Peachester, Qld 4519. (07) 5494 9712. Sleepy cod.

**Mid West Yabby and Fish Traders**, 5 Cook Dve, Swan Bay, NSW 2324. (02) 4997 5160. Bass, catfish, golden perch, silver perch, yabbies.

**Murray Cod Hatcheries**, RMB 626 Sturt Highway, Wagga Wagga, NSW 2650. (02) 6922 7360. Catfish, golden perch, Murray cod, silver perch, yabbies.

**Murray Darling Fisheries**, 1795 Old Narrandera Rd, Wagga Wagga, NSW 2650. (02) 6922 9447. Golden perch, silver perch.

**Queensland Native Fish Hatchery**, 91 Hatchery Rd, Childers, Qld 4660. (07) 4126 1844. Bass, catfish, golden perch, jade perch, sooty grunter.

**Snowy Range Aquaculture**, PO Box 316, Huonville, Tas 7109. Rainbow trout.

**South Australian Native Fresh Water Fish Hatcheries**, 31 Jikara Drive, Glen Osmond, SA 5064. (08) 8379 2463. Silver perch.

**South East Queensland Fish Breeders**, 1044 Beaudesert-Beenleigh Rd, Luscombe, Qld 4207. (07) 5546 4462. Golden perch, silver perch.

**Springfield Fisheries**, PO Box 390, Scottsdale, Tas 7260. (03) 6352 7339. Rainbow trout. <http://www.springfish.com.au>

**Sunrise Fish Farm and Hatchery**, 1359 Pacific Hwy, Kundabung, NSW 2441. (02) 6561 5133. Golden perch, silver perch.

**Sunshine Coast Crayfish Producers**, PS 1197, Yandina, Qld 4561. (07) 5446 6319. Bass, golden perch, redclaw, silver perch.

**Uarah Fish Hatchery**, Old Wagga Rd, Grong Grong, NSW 2652. (02) 6956 2245. Golden perch, silver perch.

**Wartook Native Fish Culture**, RMB 7389A Roses Gap Rd, Wartook, Vic 3401. (03) 5383 6306. Catfish, golden perch, silver perch. <http://users.netwit.net.au/~wnfc/index.html>

**WBA Hatcheries**, 2 Hamra Ave, West Beach, SA 5024. (08) 8235 0489. Barramundi.

**New Zealand Clearwater Crayfish (Koura) Ltd**, PO Box 1260, Nelson, New Zealand. (03) 546 7103. <http://www.clearwatercrayfish.co.nz/koura.html>

# Glossary

**Aeration** adding oxygen to water by churning or running air bubbles through it.

**Algae** primitive plants, in freshwater mostly relatively small and found as strands and clumps of cells, or as free-floating single-celled forms which colour water green.

**Ammonia** toxic soluble waste produced by aquatic animals, mostly present as the less-toxic ammonium, which is directly used by aquatic plants or broken down into other soluble chemicals by certain bacteria. Ammonia and ammonium are often treated as synonymous for most practical purposes.

**Buffering** adding lime to help stabilise changes in pH, so that water does not become acidic too rapidly.

**Catchment** the area over which water runs when the ground is saturated, to be collected in a pond or dam.

**Culling** removing animals which are either unhealthy or too stunted to be worth raising.

**Extensive aquaculture** raising aquatic animals in a more-or-less natural way, with no supplementary feeding and often no aeration.

**Filtration** removal of solid and dissolved wastes from water, or processing these into less harmful forms.

**Fingerling** a small, partly grown fish, similar in general appearance to the adult it will develop into.

**Fish-out** a tourist-related business raising fish for fishing, usually with a small fee for admission and the fish caught charged for by weight.

**Fry** a recently hatched or very young fish, often quite different in appearance to the adults of the species.

**Grading** separating animals into similar size or quality groups, sometimes to prevent larger ones eating the smaller ones, and sometimes for uniform presentation for sale.

**Grow-out** raising aquatic animals to a marketable size for sale.

**Intensive aquaculture** raising animals in heavily stocked conditions with the aid of filtration, aeration and regular water changes.

**Nitrate** the relatively harmless end product of the nitrogen cycle, formed by bacterial oxidation of nitrite.

**Nitrite** a toxic chemical formed by bacterial oxidation of ammonium, and broken down in turn by other bacteria to form nitrate.

**Off-flavour** an unpleasant flavour in edible aquatic animals, the cause unknown but often associated with muddy dams or raising in intensive systems. Cured by purging.

**Phytoplankton** green-water organisms including algae, mostly single-celled, which are fed upon by zooplankton.

**Plankton** free-floating or slowly swimming microscopic animals and plants.

**Purging** emptying the gut contents of aquatic animals by a period of fasting in clean water, also used to cure off-flavour.

**Siblings** variously used to mean brothers and sisters or, more loosely, just animals of a single generation.

**Terrestrial** in this book used to mean anything grown on or living on land, rather than in water or at its edge.

**Zooplankton** small, floating animals which feed on phytoplankton, in turn forming an important food for fish fry and often larger fishes.

# Index

NOTE Page numbers of illustrations are shown in **bold**.